Connected Media in the Future Internet Era

Ahmet Kondoz • Tasos Dagiuklas
Editors

Connected Media in the Future Internet Era

 Springer

Editors
Ahmet Kondoz
Institute for Digital Technologies
Loughborough University London
London, UK

Tasos Dagiuklas
Division of Computer Science
  and Informatics
London South Bank University
London, UK

ISBN 978-1-4939-8157-1   ISBN 978-1-4939-4026-4   (eBook)
DOI 10.1007/978-1-4939-4026-4

This Springer imprint is published by Springer Nature
The registered company is Springer Science+Business Media LLC New York

Contents

Chapter 1
Introduction

Ahmet Kondoz and Tasos Dagiuklas

Immersive media has received significant research interest over the last few years. As technologies have advanced dramatically in the past years, telecommunication networks can support immersive remote interaction similar to the face-to-face experience. Using immersive three-dimensional (3D) communication services, the users would look and feel as if they are physically at the same place. However, the delivery of live high-quality 3D multi-view video (MVV) to consumers remains a demanding challenge that needs efficient video coding, delivering mechanisms, and utilization of network resources [1].

By 2020, users are expected to have all of the enabling device technologies to capture, edit, process, and share high-quality content with ease, anytime anywhere with anybody. With the new content formats and capturing technologies of high-definition (HD) and ultrahigh-definition (UHD) video (including 3D), the amount and location of generated/downloaded content (since people will be moving, with such capabilities of high-quality content capture and share), and service characteristic requirements (uploading, downloading, or both), users will vary dynamically with time; for example, there will be three types of users per given time: first, passive users, downloading variable amount of content; second, those users who will capture, process, and share content by way of uploading; and finally the third group of users who will be those uploading and downloading at the same time (which may require very low latency) especially when using social networks which are

A. Kondoz
Loughborough University London, London, UK
e-mail: A.Kondoz@lboro.ac.uk

T. Dagiuklas (✉)
London South Bank University, London, UK
e-mail: tdagiuklas@lsbu.ac.uk

© Springer Science+Business Media New York 2017
A. Kondoz, T. Dagiuklas (eds.), *Connected Media in the Future Internet Era*,
DOI 10.1007/978-1-4939-4026-4_1

increasing exponentially in the number of members. One common characteristic between all three types of users will be that the amount of data transfer will be large due to the higher resolution content and usage patterns.

Advanced multimedia codecs, such as high-efficiency video coding (HEVC) [2] and its scalable and 3D video extensions, for applications such as UHD (4K) and free viewpoint TV (FTV) [3], pose new challenges in terms of compression efficiency and complexity and bit rate management.

Hence, in this evolving Future Media Ecosystem, it is normal for citizens to demand the kind of experiences that they are accustomed to in their daily/real lives, i.e., not only interact with only their own media in a "user-centric" approach, but be also able to form groups in an ad hoc manner and share their experiences in a "community-centric" approach. Overall, a community-centric media ecosystem emerges as a cross-breeding of communications, multimedia, and social network technologies, where citizens can create, share, or even trade their experiences (be called "media events"), within different contexts (e.g., home, business, mobile) and with maximum possible (i.e., better-than-best-effort) quality of experience (QoE).

HEVC is currently considered as the newest and most efficient video coding standard by the ITU-T Video Coding Experts Group (VCEG) and the ISO/IEC Moving Picture Experts Group (MPEG) [4]. It is a next-generation video coding standard that is intended for online distribution of 4K video, super high-vision broadcasting, and distribution of full HD video for mobile devices. Many researchers worldwide are looking ways to improve rate-distortion, error resiliency, and quality evaluation. Additionally, scalable extension of HEVC has recently reported in [5].

Unlike 2D, 3D poses significant challenges in terms of 3D representation (multi-view, 3D-in-2D, color plus depth, multi-view plus depth, etc.), bandwidth demands, and perceptual quality understanding [6]. 3D-HEVC is an extension of the latest compression standard HEVC (under standardization) for 3D video applications [7]. It introduces hierarchical layers and new depth coding tools. 3D-HEVC uses inter-, intra-, and inter-view predictions. A view is a combination of a texture video and its associated depth map and texture videos and depth maps are each called a layer.

ISO/MPEG is currently developing the MPEG Media Transport (MMT) as the next-generation media transport standard. MMT working draft is required to provide efficient mechanisms for delivering emerging applications and contents including 3D video, ultrahigh-definition content, multi-device, and interactive services, adaptively over All-IP, and broadcast (terrestrial, satellite, and cable) networks. The MMT specifies five different functional areas, composition: encapsulation, error control, delivery, and signalling that enable the multimedia delivery services [8].

Apart from the adoption of novel transport protocols such as MMT, adaptive streaming has received particular attention mainly driven by new deployments over the existing Web infrastructure based on HTTP. In particular, Dynamic Adaptive Streaming over HTTP (DASH) has recently been standardized by MPEG and Third Generation Partnership Project (3GPP), as a converged format for video streaming, and the standard has been adopted by other organizations fora [9].

Having access to fast broadband is important for both individuals and businesses. However, due to topographical barriers, greater geographical distances, and relatively low population density, broadband services are more difficult to obtain in rural areas and remote locations. This leads to having the issue of slow-spots and not-spots. A slow-spot is defined as the case in which the speed that can be received is less than 2 Mbps. Currently, 11 % of UK users (two million households) cannot get 2 Mbps and are in slow-spots. A not-spot is defined as the case where land-based broadband cannot be obtained at all, or only at speeds of less than 0.144 Mbps. There are 160,000 UK households that cannot receive broadband at all, or at a reasonable cost, and are therefore in a not-spot [10].

The telecommunications world today features a variety of broadband access network technologies, be it wired, wireless fixed/mobile, and various others, such as satellite networks [11]. Each of these access networks was designed for specific services and applications as well as having different capabilities in terms of bandwidth ranges, QoS support, transport mechanisms, mobility handling, and coverage. An end user having a subscription with an operator that offers services using any of these access network technologies is limited to the utilized access network's technical capabilities. Consequently, it would be desirable to have the capability to dynamically route the different users' applications and traffic over the most optimum type of access network at any time and location. Selecting which transmission link/s to use for transferring the end users' traffic at any given time and location is a challenging problem that must take into account many factors. Such factors include the type of an application (e.g., real-time, near real-time, interactive, non real-time/best-effort), its requirements on the network (e.g., throughput, latency, jitter), the type of the access network technology and its capabilities, the current network conditions (e.g., available bandwidth, congestion level, signal quality), and the end users' preferences.

Current broadband delivery is provided via ADSL, FTTK (fiber to the street plus copper to the house using VDSL), or cable modems in urban and some suburban areas. Outside of these areas it is delivered either by fixed radio extensions, by 3G/4G mobile, or by satellite which has a high latency round trip time (RTT) [11]. The EU digital agenda targets at least 30 Mbps to all households by 2020 and 100 Mbps to 50 % of households [12]. 4G mobile is gradually rolling out in Europe but it is patchy and it is estimated that by 2020 more than 50 % of households will lack connection to super-fast broadband by any of these means [13]. Extension of the 4G network and broadband into the extremes of the coverage area and to isolated communities is of key importance to the delivery of future Internet services and wealth creation. At present, satellites operating at Ka band can deliver up to 30 Mbps in the downlink for the complete coverage area, but are still limited in capacity and are separate systems to the terrestrial delivery options. The European Space Agency (ESA) has recently conducted two studies [14, 15] on hybrid use of satellite and terrestrial on the use of satellites to extend DSL systems, but these are one-way only. The EU FP7 project BATS [16] is studying advanced satellites to increase today's capacity from 100s Gbps to a Terabit/s for 2020 and beyond using Ka band to the user terminals and Q/V band for the gateways. This project also looks at intelligent routing, but from each individual user terminal via the hybrid use of satellite and terrestrial connections in order to improve the QoE to the user.

Backhaul of mobile networks has mainly been via point-to-point radio and via fiber links until recently. Only this year has Avanti operated the first backhaul of 3G mobile networks via satellite. Satellites have not been used to backhaul 4G networks to date and no hybrid networks have been applied to the backhaul.

Mobile wireless communication technologies continue to evolve (3G, 4G, 5G), offering greater and more sophisticated high-speed wireless data services, such as Internet access, video, online gaming, and much more to the end users. On one hand, there are huge interests from people for rich media applications, such as watching video and playing online games on their mobile devices, especially with the proliferation of smartphones, tablets, etc. According to a recent study [17], among all the mobile data traffic across the world, 75 % will be video by 2020. However, such rich multimedia applications are energy hungry and are one of the main factors that are contributing to the limited battery life on smartphones, in spite of the advances in smartphone technologies, which still have limited capacity of batteries and often require to be charged daily. On the other hand, mobile operators need a solution that can backhaul this large data efficiently and affordably, especially in large and highly populated urban areas.

As the 4G chapter closes, a new era beckons, which requires networking technology to evolve and to be ready for next-generation services and demand. We not only need to evolve the legacy system to be more competitive, but we also require new disruptive ideas to secure the 4G+/5G market and foster growth for the future. Indeed, we need to adopt a proactive stance in order to be ready for the 4G+ networks. In a holistic sense, 4G+ will evolve in several dimensions: use of advanced MIMO and beamforming to increase the cell capacity, deployment of small cells to offload the macro cells and reduce the coverage holes and use of virtualization and software-defined networking (SDN) technologies at the core network [18]. End users, equipped with diverse types of devices, will be able to connect to the best available connection in their close vicinity and use of 4G+ networking technologies.

We are moving to a new era of convergence between media consumer and content provider. This shift is due to the new methods of media consumption via sharing using technologies such as converged heterogeneous networks, new transport methods, personal/user-generated tools, and social media as well as new multiple interactions and collaboration among the end users [19]. As our lifestyles become more connected, even the passive behavior of watching television is turning into a very active process, where viewers are multitasking on their mobile devices while watching new forms. This shift poses new challenges in jointly optimizing media networking and sharing personalized content. Social TV refers to the convergence between broadcasting, media networking, and social networking [20]. This gives the capability to the end users to communicate, publish, and share content among themselves and interact with the TV program. Consumers have aggressively adopted online video services (e.g., Netflix, YouTube). As more providers, more content, and more devices become available, consumers seem ready to take full advantage. More consumers also expect to see increased use of video on laptops, tablets, and smartphones than on any other devices.

This book describes recent innovations in 3D media and technologies, with coverage of 3D media capturing, processing, encoding, and adaptation; networking aspects for 3D media; and quality of user experience. The structure of the book is the following.

Chapter 2 discusses the concept of effective sampling density and its application in free viewpoint video (FVV). The chapter presents how effective sampling density (ESD) is used as an indicator of signal distortion for a light field (LF)-based FVV system. Using ESD, different LF rendering methods and LF acquisition configurations are presented and compared. Eight well-known rendering methods with different acquisition configurations have been analyzed through ESD and simulation. The results have shown that ESD is an effective indicator of distortion that can be obtained directly from system parameters and takes into consideration both acquisition and rendering. Finally, several problems on FVV evaluation and optimization have been approached by using ESD. This has been done by analyzing the impact of depth estimation errors on ESD and optimization of ESD with respect to the camera density and ray selection complexity for a given output quality. Although this chapter focuses on the overall distortion of a LF-based FVV system, the concept is readily extended to measure the rendering quality at a specific location or part of the scene.

Chapter 3 is entitled "Visual Quality Regulated Three-Dimensional Video Coding (3-DVC)." The chapter reviews issues associated with 3D video coding in the context of improving viewing quality and experience of visual communications, broadcasting, and entertainment represented by UHD TV, three-dimensional digital video (3-DV), MVV, and FTV. It highlights an agonizing impasse in a much-needed paradigm shift for video coding design from a bitrate driven to a visual quality-driven design approach based on Shannon's entropy and rate-distortion (R-D). The key is to deliver with precision a designated visual picture quality discernible by human viewers for an intended application, maximizing visual experience. The chapter presents three key areas closely related to perceptual 3D video coding, including 3D video coding (3-DVC), perceptual video coding, and visual quality assessment and perceptual quality/distortion metric design.

Chapter 4 is entitled "Recent Advances on 3D Digital Video Coding and Transport: Standardization Framework and Systems." 3D video is emerging media extension of conventional 2D video into third dimension adding depth sensation and resolving viewing ambiguity. The chapter presents multi-view-plus-depth (MVD) visual representation and coding format that takes 3D geometry information of acquisition system in the form of distance information (depth map). The authors describe advances in multi-camera arrays and display technology that enable new applications for 3D video in advanced stereoscopic and auto-stereoscopic configurations. Additionally, ongoing work in MPEG/ITU standardization framework for 3D video is presented. This work regards extensions of HEVC high-efficiency video encoder. MV-HEVC extension allows efficient coding of multiple camera views and associated auxiliary pictures based on inter-view references in motion-compensated prediction. 3D-HEVC extensions allow joint coding of multiple views and associated depth maps for generating additional intermediate views in auto-stereoscopic configurations.

Chapter 5 is entitled "Depth from Defocus and Coded Apertures for 3D Scene Sensing." In 3D scene sensing, acquiring information about scene depth is important for the successful scene reconstruction and/or ultrarealistic visualization. Depth can be sensed by specifically dedicated sensors or can be estimated from multi-perspective images. Depending on whether it requires a dedicated illumination and power or purely relies on images taken at natural light, depth-sensing methods are categorized into active and passive methods, where both categories have their (application-specific) advantages and disadvantages. This chapter specifically addresses the problem of passive depth sensing utilizing the depth from defocus (DfD) effect. As a passive method, it is especially attractive since it does not require additional power and can be successfully embedded in power-constrained (e.g., mobile) devices.

Chapter 6 is entitled "Depth Map Coding for 3DTV Applications." The depth map video provides the geometry information that is used by the view synthesis to generate virtual views in between the transmitted camera views. The process used is based on depth-image-based rendering (DIBR) which projects the reference views onto the required virtual view. The depth maps are not displayed at the receiver but the synthesized views generated depend on the quality of these depth maps. Furthermore, the depth maps are composed of homogeneous areas and sharp edges, where the edges define the difference in depth and thus their preservation is very important. This suggests that coding of the depth maps must be performed diligently. This chapter explores the state of the art in depth map coding and provides insight into future directions.

Chapter 7 is entitled "Hybrid Broadcast-Broadband for the Delivery of 3D Video" and introduces an innovative hybrid broadcast-broadband scheme for enhanced television services, as it was envisaged, planned, and implemented by European-funded program ROMEO. The chapter discusses a scheme that combines terrestrial digital video broadcasting (DVB) services and peer-to-peer (P2P) overlay networks to a converged television system that supports the delivery of three-dimensional multi-view content to fixed, mobile, and portable users alongside with personal/user-generated tools and social media. This chapter describes an efficient and QoE-aware process for P2P overlay construction even in the challenging case of content delivery over multiple Internet service providers. Additionally, a novel packetization scheme is introduced that enables synchronization of streams that are received concurrently by the P2P and DVB networks, experiencing different delays in transmission. Finally, this chapter presents how the ROMEO prototype supports portable and mobile users by exploiting recent advances in wireless telecommunication systems and state-of-the-art content adaptation techniques.

Chapter 8 is entitled "HTTP Adaptive Multi-View Video Streaming." The chapter presents state-of-the-art HTTP streaming technology and the proposed adaptive MVV streaming method. When the user experiences streaming difficulties, it is necessary to perform adaptation. In the proposed streaming scenario, the optimum set of views that are predetermined by the server according to the overall MVV reconstruction quality constraint are truncated from the delivered MVV stream. In order to reconstruct the discarded views at high quality, the proposed method involves the calculation of low-overhead additional metadata at the server that is delivered to the receiver. The proposed adaptive 3D MVV streaming scheme is tested using the MPEG-DASH standard. Tests using the proposed adaptive technique have revealed that the utilization of the additional metadata in the view reconstruction process significantly improves the perceptual 3D video quality under adverse network conditions.

References

1. Apostolopoulos JG et al (2012) The road to immersive communication. Proc IEEE 100(4): 974–990
2. (2013) High efficiency video coding, document ITU-T Rec. H.265
3. Tanimoto M et al (2011) Free-viewpoint TV. Signal Process Mag IEEE 28(1):67–76
4. Sullivan GJ et al (2012) Overview of the high efficiency video coding (HEVC) standard. Circ Syst Video Technol IEEE Trans 22(12):1649–1668
5. Ye Y, Andrivon P (2014) The scalable extensions of HEVC for ultra-high-definition video delivery. MultiMedia IEEE 21(3):58–64
6. Chen Y, Vetro A (2014) Next-generation 3D formats with depth map support. MultiMedia IEEE 21(2):90–94
7. Tech G et al (2015) 3D-HEVC draft text 7, document JCT3V-K1001. Geneva
8. Lim Y et al (2013) MMT: an emerging MPEG standard for multimedia delivery over the internet. MultiMedia IEEE 20(1):80–85
9. Sodagar I (2011) The MPEG-DASH standard for multimedia streaming over the Internet. IEEE MultiMedia 4:62–67
10. Office of National Statistics (ONS) (2014) Internet access—households and individuals. http://www.ons.gov.uk/ons/dcp171778_373584.pdf
11. Hu Y, Victor OKL (2001) Satellite-based Internet: a tutorial. IEEE Commun Mag 39(3): 154–162
12. Digital Agenda: commission outlines action plan to boost Europe's prosperity and well-being. European Commission—IP/10/581. http://europa.eu/rapid/press-release_IP-10-581_en.htm?locale=en. Accessed 19 May 2010
13. Williamson B., Lewin D., Wood S. (2015), Fostering investment and competition in the broadband access markets of Europe: A report from ETNO, https://etno.eu/datas/publications/studies/PlumStudy2016.pdf, last Accessed June 2016
14. (2014) Satellite extension of xDSL copper networks—ESA ARTES 1 contract 40009106519/12/NL/CLP, https://artes.esa.int/projects/satellite-extension-xdsl-copper-wire-based-networks. Accessed 10 May 2016
15. (2014) Analysis of satellite downstream boost for xDSL networks (SAT4NET)—ESA ARTES 1 contract-AO/1-7059/12/NL/CLP, https://artes.esa.int/projects/sat4net. Accessed 10 May 2016
16. BATS: broadband access via integrated terrestrial & satellite systems. EU FP7 ICT project. http://www.batsproject.eu/
17. Cisco (2016) Cisco Visual Networking Index. Global mobile data traffic forecast update. 2015–2020 (white paper)
18. Jain R, Paul S (2013) Network virtualization and software defined networking for cloud computing: a survey. Commun Mag IEEE 51(11):24–31
19. Montpetit M-J, Klym N, Mirlacher T (2011) The future of IPTV. Multimedia Tools Appl 53(3):519–532
20. Chorianopoulos K, Lekakos G (2008) Introduction to social TV: enhancing the shared experience with interactive TV. Intl J Hum Comput Interact 24(2):113–120

Chapter 2
Quality Assessment, Evaluation, and Optimization of Free Viewpoint Video Systems by Using Effective Sampling Density

Hooman Shidanshidi, Farzad Safaei, and Wanqing Li

Abstract In a light field-based free viewpoint system (LF-based FVV), effective sampling density (ESD) is defined as the number of rays per unit area of the scene that has been acquired and is selected in the rendering process for reconstructing an unknown ray. The concept of ESD has been developed in last 7 years by the authors. It is shown that ESD is a tractable metric that quantifies the joint impact of the imperfections of LF acquisition and rendering. By deriving and analyzing ESD for the commonly used LF acquisition and rendering methods, it is shown that ESD is an effective indicator determined from system parameters and can be used to directly estimate output video quality without access to the ground truth. This claim is verified by extensive numerical simulations and comparison to PSNR. Furthermore, an empirical relationship between the output distortion (in PSNR) and the calculated ESD is established to allow direct assessment of the overall video distortion without an actual implementation of the system. A small-scale subjective user study is also conducted which indicates a correlation of 0.91 between ESD and perceived quality. ESD also has been applied to several problems for evaluation and optimization of FVV acquisition and rendering subsystems. This chapter summarizes an overview of the ESD and its application in evaluation and optimization of FVV systems.

2.1 Introduction

Free viewpoint video (FVV) [1, 2] aims to provide users the ability to select arbitrary views of a dynamic scene in real time. An FVV system consists of three main components: *acquisition* [3–8] that captures the scene using a number of cameras, *rendering* [9–16] that reconstructs the desired view from the acquired information, and *compression/transmission* [1, 2, 17–20] of captured or processed information. The performance, in particular the quality of the output video of an FVV system,

H. Shidanshidi (✉) • F. Safaei • W. Li
ICT Research Institute, Faculty of Engineering and Information Sciences,
University of Wollongong, Wollongong, NSW, Australia
e-mail: hooman@uow.edu.au; farzad@uow.edu.au; wanqing@uow.edu.au

© Springer Science+Business Media New York 2017   9
A. Kondoz, T. Dagiuklas (eds.), *Connected Media in the Future Internet Era*,
DOI 10.1007/978-1-4939-4026-4_2

depends on the efficacy of these components and their collaboration. While existing research studies individual components independently, this chapter presents a study on the joint performance of the acquisition and rendering components. The effect of compression is ignored.

In the past, studies of FVV are mainly based on simplified plenoptic signal [21] representation. In particular, by assuming that the viewer is outside of the scene, the 7D plenoptic signal is reduced to a 4D light field (LF) [22, 23]. LF refers to all the rays reflected from every point of the scene in all directions captured outside of the convex hull of the scene and a "sample" of LF refers to a discrete ray from the scene captured by a single pixel of cameras. Such LF representation has enabled the studies [3–6, 24] on the minimum sampling density under the assumption that the signal of the scene is band limited and a perfect rendering procedure is available. Results have shown that a very high camera density is required to acquire a light field, which would be infeasible in practice. On the other hand, reference-based measurements, such as peak-to-signal noise ratio (PSNR) and subjective tests [25] are usually used to assess the rendering component. These measurements require both the ground truth information and the output videos of the system, which may be a significant limitation in practice.

It is evident that both acquisition and rendering will contribute simultaneously to the signal distortion and hence the quality of the output video. This is particularly true for an FVV system that works in the *under-sampled regime* where the number of cameras deployed is not adequate to enable error-free reconstruction. To the best knowledge of the authors, before proposing ESD [26–28] there had not been any reported research on the joint impact of the two components on the output video quality. This chapter discusses this problem and reviews the theory of ESD and its application to estimate the signal distortion that accounts for both acquisition and rendering. Specifically, this chapter

- Covers the concept of effective sampling density (ESD) proposed by the authors in [26, 29] and employs it as an indicator of signal distortion for an LF-based FVV system. Calculation of ESD requires neither a reference/ground truth nor the actual output images/video. It can be derived from the key parameters of the acquisition and rendering components.
- Presents an analytical form of the ESD for the commonly used regular-grid camera systems and rendering algorithms.
- Provides theoretical and extensive empirical verification of ESD as an effective indicator of signal distortion.
- Compares ESD with PSNR, establishes an empirical relationship between them, and verifies the correlation between ESD and perceived quality through a subjective test.
- Demonstrates that how ESD can be employed for the evaluation and optimisation of FVV acquisition and rendering subsystems. Several research problems are discussed and it is shown that how ESD can be applied to these problems. The same framework can be used to similar evaluation and optimisation problems.

2.1.1 An Overview on the ESD Theory and Its Applications

The theory of ESD was first introduced by the authors in [26] followed by its application in evaluation and optimization of FVV acquisition and rendering subsystems in [29–32]. A comprehensive description of ESD and a framework for analytical derivation of ESD for different rendering methods can be found in [27, 28]. It is also shown that how theoretically calculated ESD can be used to empirically predict the output video quality in terms of objective signal distortion in PSNR as well as high correlation between ESD and perceived quality. Other applications of ESD include calculation of the minimum number of cameras for a regular camera grid [29], non-uniform light-field acquisition based on the scene complexity variations [30], and optimisation of acquisition and rendering subsystems [33].

One of the main problems in any FVV system analysis and design is acquisition and rendering evaluation and comparison. For any given acquisition configuration and rendering method, the ESD can be analytically calculated. To evaluate an acquisition component or a rendering method, it was shown in [27, 28] that the configuration or method with higher ESD has a better output video quality. Hence, ESD can be used as an unbiased tractable indicator to directly compare acquisition configurations and rendering methods.

Another important problem is acquisition and rendering optimization. To optimize the parameters of an acquisition system, e.g. camera density for a regular camera grid or the parameters of a rendering method, e.g. number of rays for interpolation, the optimization problem can be derived using the concept of ESD and solved numerically or analytically.

Another related problem is output video prediction and estimation from system parameters without the need for implementation and experiments. In [27, 28] it was shown that there is a high correlation between ESD and output video quality both in terms of objective signal distortion in PSNR and subjective quality perceived by users. In addition, an empirical method was proposed to map calculated ESD directly to rendering quality in PSNR. This allows predicting output video quality directly from FVV system parameters.

The mathematical framework to calculate ESD for a given FVV system and to solve problems of evaluation and optimization is fully addressed in [27, 28]. In this chapter a summary of some of these problems is given to show the applications of ESD.

The rest of the chapter is organized as follows. Section 2.2 reviews the related work. Section 2.3 analyses the acquisition and rendering components and describes in detail the concept of ESD. Section 2.4 presents the application of ESD to analyze LF systems with commonly used regular-grid cameras and rendering methods. Numerical simulation and validations are presented in Sect. 2.5. Section 2.6 presents the empirical relationship between the ESD and PSNR. Section 2.7 reports the subjective test and its correlation with ESD. In Sect. 2.8, several FVV research problems are discussed and it is shown that how ESD has been used or can be extended to address these problems. Section 2.9 concludes the chapter with remarks.

2.2   Related Work

This section provides a review of the existing approaches for evaluating LF acquisition and rendering methods.

2.2.1   Evaluation of the Acquisition Component

Light field can be expressed as a simplified four-dimensional plenoptic signal [21], first introduced by Levoy and Hanrahan [22] and Gortler et al. [23] (as Lumigraph) in mid-1990s. LF acquisition aims to sample the plenoptic signal by using limited number of cameras configured in 3D space. Several parameterisation schemes have been proposed to represent the camera configurations and the rays captured by the cameras. For instance, Levoy and Hanrahan [22] employed a regular grid of cameras and represented the rays by using their intersection points with two parallel planes/slabs defined by variables (s, t, u, v), respectively, where (s, t) represents the image plane and (u, v) represents the camera plane. The 4D space is then represented as a set of oriented lines, i.e. *rays* in 3D space. This parallel plane parameterisation has been enhanced by more complicated parameterisation schemes such as two-sphere (2SP) and sphere-plane parameterisation (SPP) [34].

Existing approaches for evaluating LF acquisition mainly focus on the minimum required sampling density for error-free signal reconstruction. Two major approaches have been adopted so far. The first one is based on plenoptic signal spectral analysis [3, 24] and, more specifically, the light-field spectral and frequency analysis [4, 5]. In this approach the spectral analysis is applied to a surface plenoptic function (SPF) representing the light rays starting from the object surface and the minimum sampling density is estimated based on the sampling theory by computing the Fourier transform of the light-field signal. However, the spectrum of a light field is usually not band limited due to non-Lambertian reflections, depth variations, and occlusions. Therefore, approximations such as the first-order approximation [3, 24] are often applied to the signal by assuming that the range of depth is limited.

The second approach is based on the view interpolation geometric analysis rather than frequency analysis. This approach is based on blurriness and ghost (shadow)-effect error measurements and elimination in rendered images. In [6] the artifact of "double image" (a geometric counterpart of spectral aliasing) is proposed to measure the ghost effect for a given acquisition configuration. This artifact is geometrically measured by calculating the intensity contribution of rays employed in interpolation. Finally, the minimum sampling density is calculated to avoid this error for all points in the scene. This approach can be used to derive the minimum sampling curve against scene depth information, showing how the adverse effect of depth estimation error can be compensated by increasing the sampling density, i.e. the number of cameras. This method is more flexible, especially for irregular capturing and rendering configurations, and leads to a more accurate and smaller sampling density compared with the first approach.

In addition to these two approaches, optical analysis by considering light field as a virtual optical imaging system is also employed in acquisition analysis [35, 36]. The original light field [22] shows that the distance between two adjacent cameras can be considered as the aperture for ray filtering. This concept is generalised in [14] by introducing a "discrete synthetic aperture", encompassing of several cameras. It is also shown in [14] that the size of this synthetic aperture can change the field of view very similar to an analog aperture. This optical analysis is mostly used to calculate the optimum light-field filtering [37].

Due to the assumption of perfect signal reconstruction, all of these approaches result in very high sampling densities, which are hardly achievable in practice. For instance [3] shows that for a typical scenario a camera grid with more than 10,000 cameras is required. They also assume general Whittaker–Shannon interpolation method for signal reconstruction. However, having some geometric information about the scene, such as estimated depth map, could enable more sophisticated interpolation for signal reconstruction and rendering. Consequently, an indicator to measure signal distortion without any reference or ground truth that works in the *under-sampled regime* is desirable.

2.2.2 Evaluation of the Rendering Methods

Along with the acquisition configuration and parameterisation schemes, different LF rendering methods have been developed to generate images for arbitrary viewpoints from the captured rays by implicitly or explicitly using geometric information about the scene [38]. These include layered light field [9], surface light field [10], scam light field [11], pop-up light field [12], all-in-focused light field [13], and dynamic reparameterised light field [14].

Previous works on FVV evaluation and quality assessment with respect to rendering are mainly based on the methods proposed for image-based rendering (IBR) and are not specifically for LF rendering. Often pixel-wise error metrics such as PSNR with respect to ground-truth images are employed for quality assessment [39]. Ground-truth data is provided by employing a 3D scanner for a real scene or virtual environments such as [40]. In [41], two scenarios are analysed: human performance in a studio environment and sports production in a large-scale environment. A method was introduced for both studio and large-scale environment to quantify error at the point of view synthesis [41]. This method was used as a full-reference metric to measure the fidelity of the rendered images with respect to the ground-truth as well as a no-reference metric to measure the error in rendering. In the no-reference metric, without explicitly having the ground truth, a virtual viewpoint is placed at the mid-point between the two cameras in a camera grid. From this viewpoint, two images are rendered, each using one set of the original cameras. These images are then compared against each other with the same metrics as before.

Quality evaluation has also been carried out with two different categories of metrics, modelling the human visual system (HVS) and employing more direct pixel fidelity indicators. HVS-based measures of the fidelity of an image include a variety of techniques such as measuring mutual information in the wavelet domain [42], contrast perception modelling [43], and modelling the contrast gain control of the HVS [44]. However, HVS techniques and objective evaluation of a visual system are not able to fully model the human perception as discussed in [45–47]. Pixel-wise fidelity metrics such as MSE and PSNR are simple fidelity indicators but with a low correlation with visual quality [48]. In [49] a full review of pixel-wise fidelity metrics is discussed. Also [50] shows a statistical analysis of pixel metrics and HVS-based metrics.

While the need for analytical quality evaluation of FVV systems is highlighted in several studies such as [51, 52], the current research on LF rendering evaluation and quality assessment focuses mostly on case-based study of applying these metrics. Little development has been reported on an analytical model that can evaluate LF rendering methods. In contrast, the proposed ESD provides an analytical evaluation of the effect of LF rendering as well as LF acquisition on the final video distortion.

2.3   Effective Sampling Density

Figure 2.1 shows a general FVV system that utilizes depth information. The light field is sampled by multiple cameras through the *ray capturing* process, which results in a certain sampling density (SD). SD at a given location is defined as the number of rays acquired per unit area of the convex hull of the surface of the

Fig. 2.1 The schematic diagram of a typical LF-based FVV system that utilises scene geometric information *G*

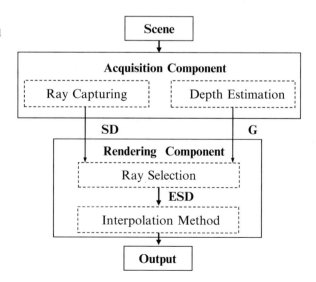

scene in that location. The acquisition can have a variety of configurations, such as regular/irregular 2D or 3D camera grids or even a set of mobile cameras at random positions and orientations. In addition, the *depth estimation* process provides an estimation of depth (e.g. depth map) to improve rendering. This could be obtained by specialised hardware, such as depth cameras, or computed from the images obtained by multiple cameras. In either case, the depth estimation will have some error.

To estimate/reconstruct an unknown ray r from the acquired rays and the depth information, the rendering essentially goes through two processes: (1) the *ray selection* that chooses a subset of acquired rays, purported to be in the vicinity of r, for the purpose of interpolation, and (2) the *interpolation* that provides an estimate of r from these selected rays.

The *ray selection process*, in particular, is often prone to error. For example, imperfect knowledge of depth may cause this process to miss some neighbouring rays and choose others that are indeed sub-optimal (with respect to proximity to r) for interpolation. Consider the case shown in Fig. 2.2, where the actual surface is at depth d and the unknown ray r intercepts the object at point p. There are four rays r_1, r_2, r_3, and r_4 captured by the cameras that lie within the interpolation neighbourhood of p, shown as a solid rectangle, and could be used to estimate r. However, since the estimation of depth is in error by Δd, the algorithm would select four other rays, r_1', r_2', r_3', and r_4', as the closest candidates for interpolation. As a result, the sampling density has been effectively reduced from $4/A$ to $4/A'$, where A and A' are the areas of solid and dashed rectangles in the figure, respectively. In addition, the rendering algorithm may not be able to use all available rays for interpolation due to computational constraint.

The output of this process, therefore, represents an *effective sampling density* (ESD) which is *lower* than the SD obtained by the cameras and distortion is inevitably introduced in the reconstructed video. ESD is defined as the number of rays per unit area of the scene that have been captured by *acquisition* component and

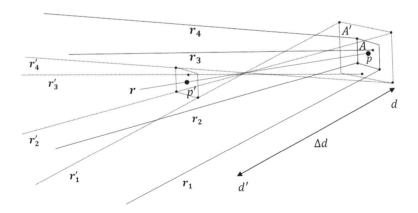

Fig. 2.2 Selection of rays in an LF rendering and the concept of ESD

chosen by *ray selection process* to be employed in the rendering. Clearly, ESD $\leq$ SD with equality holding only when the rendering process has perfect knowledge of depth and sufficient computational resources. Not surprisingly, ESD can be a true indicator of output quality, *not* SD, and its key advantage is that it provides an analytically tractable way for evaluating the influence of the imperfections of *both* acquisition and rendering components.

Let θ be the set of all rays captured by the cameras. The *ray selection mechanism* M chooses a subset ω of rays from θ. Subsequently, an *interpolation function F* is applied to ω to estimate the value of the unknown ray r. A is an imaginary convex hull area around p which intersects with all the rays in ω at depth d. The size of A would depend on the choice of ω, hence the rendering method. Note that each squared pixel in an image sensor integrates light rays coming within a squared-based pyramid extending towards the scene. The cut area (square) of this pyramid at distance d is roughly $ld \times ld$, where l is the size of the pixel determined by camera resolution. Therefore, the minimum length of the sides of A is ld, which is referred to as the system resolution in this chapter.

There are usually more rays from θ passing through A, but are not selected by the ray selection process probably because of limited computing resources or real-time requirement. Let all the captured rays passing through A be denoted by Ω. Clearly:

$$\omega \subseteq \Omega \subseteq \theta \tag{2.1}$$

Both M and F may or may not use some kind of scene geometric information G such as focusing depth (average depth of the scene computed from automatic focusing algorithms or camera distance sensors) or depth map. Mathematically, the rendering can be formulated as

$$\omega = M(\theta, G) \tag{2.2}$$

$$r = F(\omega, G) \tag{2.3}$$

Different rendering methods differ in their respective M and F functions and their auxiliary information G.

Based on these definitions SD and ESD can be expressed as

$$SD = \frac{|\Omega|}{A} \tag{2.4}$$

$$ESD = \frac{|\omega|}{A} = \frac{|M(\theta, G)|}{A} \tag{2.5}$$

where $|\Omega|$ and $|\omega|$ are the number of rays in Ω and ω, respectively. A is the area of interpolation convex hull, and can be calculated by deriving the line equations for the boundary rays β_i's and finding the vertexes of convex hull A at depth d. Figure 2.3 shows this process for a simple 2D LF acquisition, generated by applying

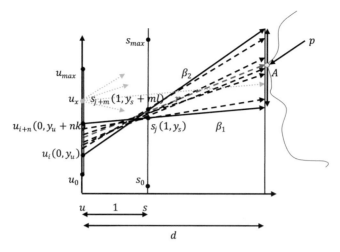

Fig. 2.3 ESD calculation for a simplified 2D light-field system

a 2D projection to a 3D light field with two-plane parameterisation, that is, camera plane uv and image plane st over (u, s). Assume that rays in ω are surrounded by the boundary rays β_1 and β_2. The rays in ω are selected by the selection method M and are bounded by $n + 1$ cameras in u (u_i to u_{i+n}) and $m + 1$ pixels in s (s_j to s_{j+m}). As it can be seen, A is at least a function of k, l, n, m, and d, where k is the distance between the cameras, l is the pixel length, n and m are the number of cameras and pixels bounded by boundary rays, respectively, and d is the depth of p.

The rays that intersect with A from these $n + 1$ cameras are the rays employed by rendering method, i.e. ω set. However, as it is shown in Fig. 2.3, there are more than $n + 1$ cameras in the grid (in addition to cameras bounded between u_i and u_{i+n}) that are able to see area A. u_x is shown as an example of these cameras. The rays from these cameras to A make up the difference between Ω and ω sets.

SD defined in Eq. (2.4) provides the upper bound of ESD. In general, for a given LF acquisition configuration, it is possible to calculate SD on any point over the scene space analytically or numerically. SD is generally not uniform across the field of view, even when a regular camera grid is used in capturing. Figure 2.4a shows the SD contour maps at different depths, $d = 30$, 60, and 90 m, for a regular camera grid of 30×30 with $k = 2$ m, camera field of view of $30°$, image resolution of 100×100 pixels, i.e. $l = 0.53$ cm in image plane st, and ideal area $A = (ld)^2$, i.e. LF system resolution. Figure 2.4b shows a 2D slice where d ranges in [2 m, 100 m].

Based on the discussion above, it can be speculated that the output quality of an arbitrary view is determined by three key factors: ESD in each area A, the vicinity of the unknown rays that compose the view, scene complexity in each area A, which could be measured in terms of its spatial frequency components, and the interpolation function F employed for the estimation of the unknown rays.

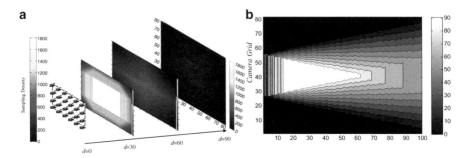

Fig. 2.4 (**a**) SD contour maps at different depths in 3D; (**b**) SD contour map in 2D

In particular, for a fixed scene complexity and a given interpolation algorithm, ESD can be used to analytically estimate the signal distortion of a given camera configuration and an adopted rendering algorithm.

2.4 ESD Analysis of LF Rendering Methods

Without loss of generality, a simple regular-grid camera system, as shown in Fig. 2.3, is adopted in this section. ESD analysis is presented for different rendering algorithms, specifically those with and without using depth information. However, the analysis can be extended to other acquisition systems [34]. For a regular-grid camera system, analytical form of ESD can be obtained for a rendering algorithm with and without using depth information.

2.4.1 Rendering Methods Without the Depth Information

The LF rendering methods without using depth information, hereafter referred to as *blind* methods, can be categorised into four main groups based on their ray selection mechanism M: nearest neighbourhood estimation (NN), 2D interpolation in camera plane (UV), 2D interpolation in image plane (ST), and a full 4D interpolation in both camera and image planes (UVST) [22, 53]. For interpolation function F, bilinear interpolation is often used for the 2D interpolation and a quadrilinear interpolation for the 4D interpolation. However, when $|\omega| > 4$ for UV and ST and when $|\omega| > 16$ for UVST, the convex hull A may not be a grid anymore and other types of 2D and 4D interpolation function F could be employed. This will be discussed later in subsection 2.4.3.

Considering the regular geometry of the cameras shown in Fig. 2.3, analytical form of ESD for these rendering algorithms can be derived. Table 2.1 summarises the ESD derivation for the NN, ST, UV, and UVST methods where $|\omega| = 4$ for UV

Table 2.1 ESD for the LF rendering methods without using depth information [25]

Rendering method	Selection mechanism M	Interpolation function F	Sampling/interpolation length A in 2D LF	ESD for symmetric 3D light field
NN	Select the nearest ray in 4D space, $|\omega| = 1$	No interpolation, neighbourhood estimation	$A_{NN} = \left(\frac{1+k}{2}\right) d - \frac{k}{2}$	$ESD_{NN} = \frac{1}{A_{NN}^2}$
ST	Select four or more rays from the neighbourhood pixels in st plane to the nearest camera in uv plane, $|\omega| \geq 4$	Any type of 2D interpolation, e.g. bilinear interpolation for 2D grid selection of rays	$A_{ST} = \left(l + \frac{k}{2}\right) d - \frac{k}{2}$	$ESD_{ST} = \frac{4}{A_{ST}^2}$
UV	Select four or more rays from the neighbourhood cameras in uv plane to the nearest pixel in the st plane, $|\omega| \geq 4$	Any type of 2D interpolation, e.g. bilinear interpolation for 2D grid selection of rays	$A_{UV} = \left(k + \frac{l}{2}\right) d - k$	$ESD_{UVST} = \frac{4}{A_{UVST}^2}$
UVST	Select 16 or more rays from four neighbourhood cameras in uv to 4 neighbourhood pixels in st, $|\omega| \geq 16$	Any type of 4D interpolation, e.g. quadrilinear interpolation for grid selection of rays	$A_{UVST} = (l + k) d - k$	$ESD_{UVST} = \frac{16}{A_{UVST}^2}$

and ST and $|\omega| = 16$ for UVST. For each one of these rendering methods, the details of selection mechanism M and interpolation function F are given in the second and third columns. The fourth column summarises the sampling/interpolation length A. Notice that A is a segment in the chosen 2D LF system whereas it is an area in 3D. The fifth column lists the corresponding ESD.

With the analytical ESD forms shown in Table 2.1, it is possible to objectively compare these rendering methods in terms of the signal distortion for the same acquisition. The higher the ESD is, the less distortion is expected. Since when $|\omega|$ is fixed, ESD is a function of the sampling/interpolation area A. The ratio γ of A between two rendering methods is used as a factor for comparison.

Table 2.2 summarises the comparison. The first column shows a pair of rendering methods to be compared, the second column is the ratio γ, the third column gives the relationship between the corresponding ESDs, and the fourth column is the minimum value of γ for each pair. Specifically, three particular scenarios are analysed and their corresponding γ are shown in the fifth column of Table 2.2.

Scenario One: $d \to \infty$ and $k \gg l$, which represents a typical low-density camera grid and a scene that is very far from the cameras. In this case, the analysis shows that $4\text{ESD}_{NN} < 4\text{ESD}_{UV} < \text{ESD}_{ST} < \text{ESD}_{UVST}$. In other words, UVST has the highest ESD and is expected to produce the video with least distortion. NN has the lowest ESD and therefore would generate output with a larger distortion.

Scenario Two: $d \to \infty$ and $k \cong l$, a hypothetical very-high-density camera grid for a scene that is very far from the grid. The analysis indicates that $1.7\text{ESD}_{NN} < \text{ESD}_{UV} < \text{ESD}_{ST}$, $4\text{ESD}_{NN} < \text{ESD}_{UVST}$, and $2.2\text{ESD}_{UV} < 2.2\text{ESD}_{ST} < \text{ESD}_{UVST}$. This shows the same order as first scenario, but both NN and UV methods work much better in comparison with ST, though UVST still has the best performance.

Scenario Three: $d \cong 1$, a hypothetical scene very close to the image plane. The analysis indicates that $4\text{ESD}_{NN} < 4\text{ESD}_{ST} < \text{ESD}_{UV} < \text{ESD}_{UVST}$. This shows that UV outperforms ST in such a scenario with ESD more than four times higher than ST. Hence, for a scene close to the grid, UV is a better choice for rendering method compared with ST, which is intuitively appealing.

Similar analysis can be applied to other scenarios, which can offer a choice of rendering algorithms for a given acquisition system.

2.4.2   Rendering Methods with the Depth Information

Utilisation of depth information G in rendering can compensate to some extent for insufficient number of samples acquired in an *under-sampling* situation [54]. It can make the ray selection mechanism M more effective compared with blind rendering methods. The amount of depth information G could vary from a crude estimate, such as the focusing depth, to the full depth map or even full 3D geometric model of the scene. A mechanism M in this case may choose a number of rays intersecting the scene in the vicinity of point p at depth d. A rendering method whose interpolation function F is a 2D interpolation over uv plane and utilises only the focusing depth

Table 2.2 Comparison of ESD of the LF rendering methods without using depth information [25]

Methods	Sampling length comparison	ESD comparison	γ (The ratio of ESDs)	γ Analysis
NN vs. ST	$A_{NN}\gamma > A_{ST}$	$ESD_{NN}\frac{4}{\gamma^2} < ESD_{ST}$	$\gamma > 1 + \frac{ld}{(l+k)d-k}$	$d \to \infty$ and $k \gg l \Rightarrow \gamma = 1$ $d \to \infty$ and $k \approx l \Rightarrow \gamma = 1.5$ $d \approx l \Leftarrow \gamma = 2$
NN vs. UV	$A_{NN}\gamma > A_{UV}$	$ESD_{NN}\frac{4}{\gamma^2} < ESD_{UV}$	$\gamma > 1 + \frac{kd-k}{(l+k)d-k}$	$d \to \infty$ and $k \gg l \Rightarrow \gamma = 2$ $d \to \infty$ and $k \approx l \Rightarrow \gamma = 1.5$ $d \approx l \Leftarrow \gamma = 1$
NN vs. UVST	$A_{NN}\gamma > A_{UVST}$	$ESD_{NN}\frac{16}{\gamma^2} < ESD_{UVST}$	$\gamma > 2$	$\gamma > 2$
ST vs. UVST	$A_{ST}\gamma > A_{UVST}$	$ESD_{ST}\frac{4}{\gamma^2} < ESD_{UVST}$	$\gamma > 1 + \frac{d-1}{(\frac{2l}{k}+1)d-1}$	$d \to \infty$ and $k \gg l \Rightarrow \gamma = 2$ $d \to \infty$ and $k \approx l \Rightarrow \gamma = 1.33$ $d \approx l \Leftarrow \gamma = 1$
UV vs. UVST	$A_{UV}\gamma > A_{UVST}$	$ESD_{UV}\frac{4}{\gamma^2} < ESD_{UVST}$	$\gamma > 1 + \frac{ld}{(l+2k)d-2k}$	$d \to \infty$ and $k \gg l \Rightarrow \gamma = 1$ $d \to \infty$ and $k \approx l \Rightarrow \gamma = 1.33$ $d \approx l \Leftarrow \gamma = 2$
ST vs. UV	$A_{UV} > \gamma A_{ST}$	$ESD_{UV}\gamma^2 < ESD_{ST}$	$\gamma < 1 + \frac{(k-l)d-k}{(2l+k)d-k}$	$d \to \infty$ and $k \gg l \Rightarrow \gamma = 2$ $d \to \infty$ and $k \approx l \Rightarrow \gamma = 1$ $d \approx l \Leftarrow \gamma = 0.5$

is referred to as UV-D (**UV** + **D**epth) and the one with a full depth map is referred to as UV-DM (**UV** + **D**epth **M**ap). By extending the selection mechanism M and interpolation function F to a full 4D interpolation over both uv and st planes, the rendering methods are referred to as UVST-D (**UVST** + **D**epth) and UVST-DM (**UVST** + **D**epth **M**ap), respectively, the former using focusing depth only. Many LF rendering methods with depth information can be mathematically expressed in the form of one of these four groups. These include layered light field [9], surface light field [10], scam light field [11], pop-up light field [12], all-in-focused light field [13], and dynamic reparameterised light field [14].

Again, without loss of generality, we study the cases where $|\omega| = 4$ and bilinear interpolation as F for UV-D and UV-DM and $|\omega| = 16$ and quadrilinear interpolation as F for UVST-D and UVST-DM.

Figure 2.5 illustrates the rendering methods with depth information. If the exact depth d at point p, the intersection of unknown ray r with the scene, is known, applying a back projection can find a subset of known rays Ω intersecting the scene at the vicinity of p. Subsequently, an adequate subset ω of these rays can be selected by mechanism M to be employed in interpolation F.

However, in practice, the estimated depth of p has an error Δd. This makes the rays intersect in an imaginary point p' in the space and going through the vicinity of area A on the scene instead of intersecting with the exact point p on the scene surface. Subsequently, this estimation error Δd would result in reduction of ESD and increase the distortion. To compute Ω in this case, back projection should be applied to the vertexes of A and not p to find all the rays passing through A.

The size of area A depends on Δd and as Δd gets larger it also increases. Usually only the upper bound of the error is known and therefore in this chapter the worst-

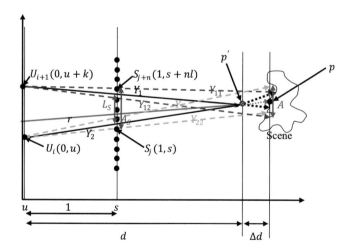

Fig. 2.5 Light-field rendering methods using depth information (UV-D, UVST-D, UV-DM/UVST-DM) with Δd error in depth estimation

case scenario, i.e. largest A, is computed in the LF analysis which corresponds to the lower bound of ESD.

Considering scenario in Fig. 2.5, Y_1 and Y_2 are two immediate-neighbour rays, intersecting with the desired ray r at depth d on object surface. If these two rays don't pass through the known s values in image plane, Y_1 from Y_{11} and Y_{12} and Y_2 from Y_{21} and Y_{22} can be estimated. Finally, a bilinear interpolation in uv plane (or a linear interpolation over u in this 2D example) is applied to estimate r from Y_1 and Y_2.

Here, ω includes only two samples for UV-D/UV-DM and four samples for UVST-D/UVST-DM though all acquired rays that intersect the object surface at point p in vicinity A at depth d can be employed in the rendering ($\omega = \Omega$) to reduce distortion. Y_{12} and Y_{21} are boundary rays used for interpolation. If the depth estimation has no error, i.e. $\Delta d = 0$, then $A_S = L_S + \frac{l}{2} + \frac{l}{2} = \frac{k(d-1)+ld}{d}$, $A_{UVD/UVDM} = ld$, and $A_{UVSTD/UVSTDM} = 2ld$. In a case that $\Delta d > 0$, p is somewhere in the range of $d \pm \Delta d$, and the sampling area A would be increased to

$$A = \max\left[\left|Y_{11}\left(d + \Delta d\right) - Y_{22}\left(d + \Delta d\right)\right|, \left|Y_{12}\left(d + \Delta d\right) - Y_{21}\left(d + \Delta d\right)\right|\right]$$

$$= l\left(d + \Delta d\right) + \frac{\Delta d \times k}{d} \tag{2.6}$$

Using this approach, it can be shown that the difference between the rendering methods with focusing depth (UV-D/UVST-D) and the rendering methods with full depth map (UV-DM/UVST-DM) is in the scale of Δd. For focusing depth, a fixed depth is used for all points of the scene. This makes the depth estimation error $\Delta d = \frac{object\ length}{2} + focusing\ depth\ estimation\ error$. When the full depth map of the scene is used as G, the depth of each point p of the scene possibly with some estimation error Δd is known. Δd is usually much less than the focusing depth error, which makes the UV-DM/UVST-DM rendering less distorted than UV-D/UVST-D.

2.4.3   General Case of Rendering Methods with Depth Maps

Figure 2.6 demonstrates an LF rendering method with two-plane parameterisation using a depth map as the auxiliary information G. Again ray r is the unknown ray that needs to be estimated for an arbitrary viewpoint reconstruction. r is assumed to intersect the scene on point p at depth d.

In Fig. 2.6, seven rays from all rays intersecting imaginary p are selected by M, i.e. $|\omega| = 7$, assuming that these rays pass through known pixel values or if neighbourhood estimation is used. In the case of bilinear interpolation in st plane, 28 rays are chosen by M to estimate these 7 rays. The chosen cameras in uv plane are bounded by a convex hull A'. It is easy to show that interpolation convex hull A is proportional to A'.

Finally a 2D interpolation F over convex hull A' on uv plane can be applied to estimate unknown ray r from the rays in ω. This rendering method with depth information is a generalisation of UV-DM described in subsection 2.4.2 but with

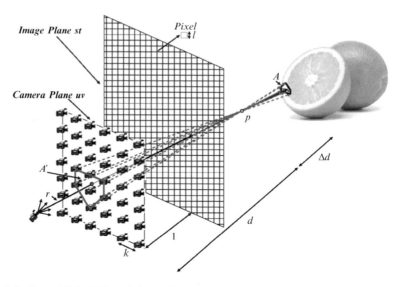

Fig. 2.6 General light-field rendering method using depth information (UV-DM/UVST-DM) with Δd error in depth estimation

arbitrary number of rays for interpolation when 2D interpolation is performed over neighbouring cameras in the *uv* plane and neighbourhood estimation, i.e. choosing the closest pixel in the *st* plane. Again the generalisation of UVST-DM is in the case of 2D interpolation over neighbouring cameras in the *uv* plane and bilinear interpolation over neighbouring pixels in the *st* plane.

In a simple form of UV-DM and UVST-DM, the rays in ω are selected in a way that A' becomes rectangular, i.e. 2D grid selection and therefore 2D interpolation over A' can be converted into a familiar bilinear interpolation.

The ESD for the UV-DM and UVST-DM demonstrated in Fig. 2.6 can be derived as

$$\text{ESD}_{\text{UVDM}} = \frac{|\omega|}{A} = \frac{|\omega|}{\frac{\Delta d}{d}A' + \mu\left(l\left(d + \Delta d\right), A'\right)} \tag{2.7}$$

$$\text{ESD}_{\text{UVSTDM}} = \frac{|\omega|}{A} = \frac{|\omega|}{\frac{\Delta d}{d}A' + \mu\left(2l\left(d + \Delta d\right), A'\right)} \tag{2.8}$$

where μ is a function to calculate the effect of pixel interpolation over *st* plane on the area A. A is mainly determined by A', but the pixel interpolation μ which is added to Eqs. (2.7) and (2.8) also has a small effect on A. The pixel interpolation over *st* even when $\Delta d = 0$ makes $A = (ld)^2$.

Simple forms of UV-DM and UVST-DM described in subsection 2.4.2 can be formulated for a regular camera grid and 2D grid selection of rays, i.e. A' as a rectangular area with 4 and 16 samples in $|\omega|$, respectively; subsequently Eqs. (2.7) and (2.8) become

$$\text{ESD}_{\text{UVDM}} = \frac{4}{\left(\frac{\Delta d.k}{d} + l\,(d + \Delta d)\right)^2} \tag{2.9}$$

$$\text{ESD}_{\text{UVSTDM}} = \frac{16}{\left(\frac{\Delta d \times k}{d} + 2l\,(d + \Delta d)\right)^2} \tag{2.10}$$

where k is the distance between the two neighbouring cameras in the camera grid and l is the length of the pixel in the image plane as illustrated in Fig. 2.6. Note that the edge of rectangular A' is equal to k and that is how Eqs. (2.9) and (2.10) are derived from Eqs. (2.7) and (2.8).

Mathematically, a general representation of simplified UV-DM rendering method with arbitrary number of rays for interpolation is $r = \text{UVDM}\,(d, \Delta d, k, l, |\omega|)$. By extending Eq. (2.9) and considering the edge of rectangular A' to be equal to $\left(\sqrt{|\omega|} - 1\right) k$, the ESD could be calculated for $\text{UVDM}(d, \Delta d, k, l, |\omega|)$ as follows:

$$\text{ESD}_{\text{UVDM}(d,\Delta d,k,l,|\omega|)} = \frac{|\omega|}{\left(l\,(d + \Delta d) + \frac{\Delta d \times k}{d}\left(\sqrt{|\omega|} - 1\right)\right)^2} \tag{2.11}$$

Equation (2.11) assumes that the rays are chosen for interpolation symmetrically around the vertical and horizontal axes, such as 4×4 samples. In this case, $\sqrt{|\omega|}$ would be an integer.

ESD for the rendering methods using either focusing depth or depth maps can be analytically derived based on the geometry of the regular grid camera system as described in Figs. 2.5 and 2.6, Eqs. (2.7), (2.8), (2.9), (2.10), and (2.11). Table 2.3 summarises derivation. The first column shows the rendering methods: UV-D and UVST-D methods that use focusing depth and UV-DM and UVST-DM that use depth maps, with $|\omega| = 4$ or 16 and $|\omega| > 4$ or 16. The second and third columns describe the selection mechanism M and interpolation function F, respectively. The fourth and fifth columns give the sampling/interpolation length A and ESD, respectively.

Table 2.4 summarises comparison of the ESD among UVST, UV-D, and UVST-D. It is clear from Table 2.3 that (UV-DM and UV-D) and (UVST-DM and UVST-D) have the same ESD, the difference between them being the scale of Δd; thus UV-DM and UVST-DM are omitted in Table 2.4. Similar to the analysis of the blind methods, ratio γ is used and two scenarios, one with $d \to \infty$, $k \cong l$ and $\Delta d \ll d$ and the other with $d \to \infty$, $k \gg l$ and $\Delta d \ll d$, are analysed. The second scenario corresponds to a typical FVV system where the scene is far from the camera grid, depth estimation error is small compared with the depth, and there are a finite number of cameras.

The γ values allow us to compare the rendering methods with and without using depth information. Tables 2.2 and 2.4 have shown that $4\text{ESD}_{\text{NN}} < 4\text{ESD}_{\text{UV}} <$

Table 2.3 ESD for the LF rendering methods with depth information

Rendering method category	Selection mechanism M	Interpolation function F	Sampling/interpolation length A in 2D LF	ESD for symmetric 3D light field
UV-D $\lvert\omega\rvert = 4$	Select four rays sourcing from neighbourhood cameras in uv and intersecting with expected p	Neighbourhood estimation in st and 2D interpolation over uv	$A_{UVD} = l(d + \Delta d) + \dfrac{\Delta dk}{d}$	$ESD_{UVD} = \dfrac{4}{A_{UVD}^2}$
UVST-D $\lvert\omega\rvert = 16$	Select 16 rays sourcing from neighbourhood cameras in uv, through known pixels in st and intersecting with expected p	4D interpolation over st and uv planes, e.g. quadrilinear interpolation	$A_{UVSTD} = 2l(d + \Delta d) + \dfrac{\Delta dk}{d}$	$ESD_{UVSTD} = \dfrac{4}{A_{UVSTD}^2}$
UV-DM $\lvert\omega\rvert = 4$	The same as UV-D but with more accurate depth estimation of p employing depth maps.	The same as UV-D	$A_{UVDM} = l(d + \Delta d) + \dfrac{\Delta dk}{d}$	$ESD_{UVDM} = \dfrac{4}{A_{UVDM}^2}$
UVST-DM $\lvert\omega\rvert = 16$	The same as UVST-D but with more accurate depth estimation of p employing depth maps	The same as UVST-D	$A_{UVSTDM} = 2l(d + \Delta d) + \dfrac{\Delta dk}{d}$	$ESD_{UVSTDM} = \dfrac{16}{A_{UVSTDM}^2}$
UV-DM $\lvert\omega\rvert > 4$	Select $\lvert\omega\rvert$ rays sourcing from neighbourhood cameras in uv and intersecting with expected p	2D interpolation over chosen rays in ω and estimate each ray from closest known pixel in st	$A_{UVDM(d,\Delta d,k,l,\lvert\omega\rvert)} = l(d + \Delta d) + \dfrac{\Delta dk}{d}\left(\sqrt{\lvert\omega\rvert} - 1\right)$[a]	$ESD_{UVDM(d,\Delta d,k,l,\lvert\omega\rvert)} = \dfrac{\lvert\omega\rvert}{A_{UVDM(d,\Delta d,k,l,\lvert\omega\rvert)}^2}$
UVST-DM $\lvert\omega\rvert > 16$	Select $\lvert\omega\rvert$ rays sourcing from neighbourhood cameras in uv, through known pixels in st and intersecting with expected p	4D interpolation over chosen rays in ω in both uv and st planes	$A_{UVSTDM(d,\Delta d,k,l,\lvert\omega\rvert)} = 2l(d + \Delta d) + \dfrac{\Delta dk}{d}\left(\sqrt{\lvert\omega\rvert} - 1\right)$[a]	$ESD_{UVSTDM(d,\Delta d,k,l,\lvert\omega\rvert)} = \dfrac{\lvert\omega\rvert}{A_{UVSTDM(d,\Delta d,k,l,\lvert\omega\rvert)}^2}$

[a]This is calculated by assuming that chosen rays are form a rectangular grid in uv plane for simplification

Table 2.4 Comparison of the UVST, UV-D/UV-DM, and UVST-D/UVST-DM methods

Methods	Sampling length comparison	ESD comparison	γ Ratio	γ Analysis
UVST vs. UV-D	$A_{UVST} > \gamma A_{UVD}$	$ESD_{UVST}\frac{\gamma^2}{4} < ESD_{UVD}$	$\gamma < \dfrac{(k+l)d^2 - kd}{ld^2 + l\Delta dd + k\Delta d}$	$d \to \infty, k \cong l$ and $\Delta d \ll d \Rightarrow \gamma = 2$ $d \to \infty, k \gg l$ and $\Delta d \ll d \Rightarrow \gamma = \infty$
UVST vs. UVST-D	$A_{UVST} > \gamma A_{UVSTD}$	$ESD_{UVST}\gamma^2 < ESD_{UVSTD}$	$\gamma < \dfrac{(k+l)d^2 - kd}{2ld^2 + 2l\Delta dd + k\Delta d}$	$d \to \infty, k \cong l$ and $\Delta d \ll d \Rightarrow \gamma = 1$ $d \to \infty, k \gg l$ and $\Delta d \ll d \Rightarrow \gamma = \infty$
UV-D vs. UVST-D	$A_{UVD} > \gamma A_{UVSTD}$	$ESD_{UVD}4\gamma^2 < ESD_{UVSTD}$	$\gamma < 1 - \dfrac{ld^2 + l\Delta dd}{2ld^2 + 2l\Delta dd + k\Delta d}$	$d \to \infty \Rightarrow \gamma = \frac{1}{2}$

$ESD_{ST} < ESD_{UVST} \ll ESD_{UVD/UVDM} < ESD_{UVSTD/UVSTDM}$, i.e. for a given acquisition, the NN rendering method has the lowest ESD and hence results in the highest video distortion followed by UV, ST, UVST, UV-D/UV-DM, and UVST-D/UVST-DM, respectively. The experimental validation in next section will not only confirm this, but also show that ESD is highly correlated with PSNR.

Equations shown in Tables 2.3 and 2.4 can be used in LF system analysis and design. In addition to LF system evaluation and comparison, by knowing the upper bound of the depth estimation error, optimum system parameters such as camera density k, camera resolution in terms of l, and rendering complexity in terms of number of rays employed in interpolation $|\omega|$ can be theoretically calculated. For example, in [29], the authors have used the above relationships to obtain the minimum camera density for capturing a scene. We will show in future publications how ESD can be used to optimise the acquisition and rendering parameters of an LF system individually and jointly for a target output video quality.

2.5 Theoretical and Simulation Results

To verify the effectiveness of ESD as an indicator to estimate the distortion introduced by the acquisition and rendering components in an LF-based FVV system, a computer simulation system employing a 3D engine has been developed to generate the ground-truth data [55]. The system takes a 3D model of a scene and simulates a multiple camera system to capture the scene. For any virtual views to be reconstructed, the system generates its ground-truth image as a reference for comparison. Figure 2.7 illustrates a simulated regular-camera grid for acquisition. Virtual views were randomly generated as the ground truth and used to evaluate the performance of ESD as a distortion indicator.

In addition, since 3D models were used to represent the scene, a full precise depth map was available for rendering. Error is simulated and added to the depth map in order to evaluate ESD when inaccurate depth is employed in the rendering. In the following, details on the depth error model and experimental settings are presented.

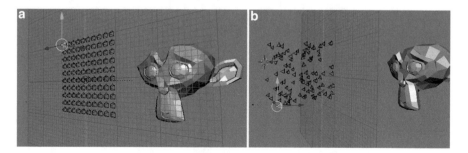

Fig. 2.7 (a) A simulated regular camera grid; (b) random virtual viewpoints

2.5.1   Depth Error Model

There are two commonly used approaches to obtain depth information for FVV systems [56]: triangularisation based through either stereoscopic vision or structure light, and time-of-flight (ToF) based. When depth is estimated using the former approach, the error Δd is normally distributed whose standard deviation is proportional to the square of distance d^2, i.e. $\Delta d \approx \tau \times d^2$, where τ depends on the system parameters [57]. For ToF, the error tends to be approximated coarsely as $\Delta d \approx \tau \times d$ [58]. The linear model is adopted for the experimental validation in this chapter. In the experiments, the ground-truth depth map is known from the simulator. Based on the prescribed depth estimation error, for each pixel of the exact depth map, a random error with normal distribution and standard deviation of $\Delta d = \tau \times d$ is introduced to create a noisy depth map with average of τ % error.

2.5.2   ESD of Scenes

The ESD equations summarised in Tables 2.1 and 2.3 are all for a small vicinity of scene around a given point p. Clearly, ESD varies over the scene, depending on the depth. On the other hand, the overall distortion of output in addition to ESD is also scene dependent. Estimation of overall distortion for a given scene requires integration of ESD over the entire scene and at each point considering the scene texture complexity. In this chapter, an approximation is adopted by using the average depth of the scene. This allows analysing acquisition configurations or rendering methods based on ESD independently of the scene complexity. To compare acquisition configurations and rendering methods an $\overline{\text{ESD}}$ for each configuration/method is calculated for comparison using an average depth of the scene $\bar{d}$ with an average $\overline{\Delta d}$ of absolute depth error.

2.5.3   Simulation Settings

For the experiments reported in this chapter, the LF engine is customised for the eight LF rendering methods: NN, UV, ST, UVST, UV-D, UVST-D, UV-DM, and UVST-DM with $|\omega| = 1, 4, 4, 16, 4, 16, 4, 16$, respectively, with default rectangular grid ray selection for M and bilinear and quadrilinear interpolations for F.

To assess the effect of scene complexity on output distortion, four 3D models, a "*room*", a "*chess board*", "*blender monkey*", and "*Stanford bunny*", as shown in Fig. 2.8, were selected, where the complexity decreases in this order. In the simulation, the centre of the 3D model was placed at $d = 10$ m by default, if depth is not given in the experiment. A 16×16 regular camera grid was placed for

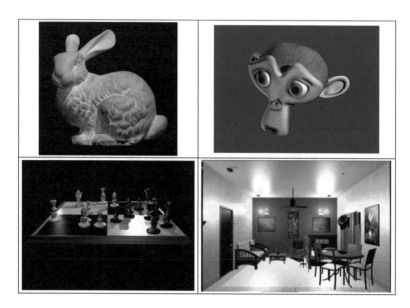

Fig. 2.8 Four 3D scenes chosen for experimental validation

acquisition and the image resolution was originally set to 1024×768 pixels, i.e. $l = 0.05$. However, for experiments reported in Fig. 2.10, to evaluate the effect of the 3D model depth in output PSNR, $\bar{d}$ is changed between [10 m, 50 m], in Fig. 2.17 to evaluate the effect of the camera grid density in output PSNR, k is changed between [0.1 m, 0.9 m], and in Fig. 2.19 to evaluate the effect of the reference camera resolution on output PSNR, l is changed between [0.02 cm, 0.1 cm], to analyse the effects of these factors on the output distortion. Please note that the term pixel size in the following experiments refers to l, the projected pixel size on image plane st at depth $d = 1$. Hence, $l = 0.02$ cm on st plane corresponds to a real pixel size equal to 4.8×10^{-4} cm for a typical $1/2''$ camera sensor or capturing resolution of 2560×1920. With the same assumptions, $l = 0.05$ cm corresponds to capturing resolution of 1024×768 and $l = 0.1$ cm to resolution of 512×384.

For each 3D model, 1000 random virtual cameras at different distances from the scene were generated and average PSNR between the rendering images and the ground truth was calculated for comparison. In the following, the theoretical expectations in terms of calculated $\overline{ESD}$ and the actual measurement of output video distortion in PSNR are reported and compared for different rendering methods and different acquisition configurations.

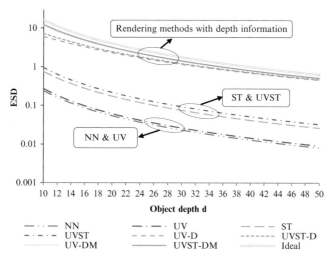

Fig. 2.9 Theoretical $\overline{ESD}$ for different LF rendering methods based on object depth $\bar{d}$ for $k = 0.4$m and $l = 0.05$cm (i.e. camera resolution of 1024×768)

2.5.4 Results on Rendering Methods

2.5.4.1 Theoretical Expectation

Figure 2.9 shows the ESD for the above-mentioned LF rendering methods in addition to the ideal rendering ($\Delta d = 0$) where $k = 0.4$ m, $l = 0.05$ cm, $d \in [10 \text{ m}, 50 \text{ m}]$, the object length is 5 m, and $\Delta d = 0.1d$, i.e. 10 % error in depth estimation. The ideal case is when there is no error in the depth map and refers to the maximum value for ESD at depth d. The vertical axis is logarithmic. For UV-D and UVST-D the actual error is $\frac{\text{object length}}{2} + \Delta d$, which in this example is equal to $2.5\text{m} + 0.1d$.

It can be seen from Fig. 2.9 that, for all depths, the expected relative relationship of ESD among the eight LF rendering methods is maintained. A quadrilinear interpolation over UVST makes UVST-D and UVST-DM perform slightly better than their corresponding UV-D and UV-DM, especially for small d. For large depths, UV-D/UVST-D performance approaches that of UV-DM/UVST-DM, because the object length is small compared to depth error in this case.

Figure 2.11 demonstrates a bar chart of theoretical ESD values for different rendering methods for $k = 0.4$ m and $l = 0.05$ cm, for a point p with $d = 10$ m and $\Delta d = 1$ m.

Figure 2.13 shows the effect of depth map error on ESD for UV-DM for $l = 0.01$ cm, $|\omega| = 4$, $\bar{d} = 100$, and $\frac{\Delta d}{d}$ between 0 % and 20 %, for $k = 5, 10, 20,$ and 50. As it can be seen, higher errors in depth estimation result in less ESD when k is fixed. However, small k could increase the ESD.

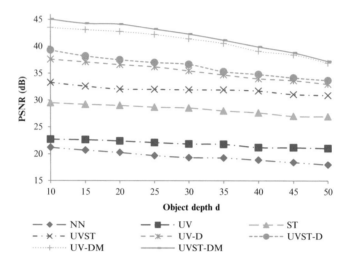

Fig. 2.10 Experimental rendering quality in PSNR for different LF rendering methods vs. object depth $\bar{d}$

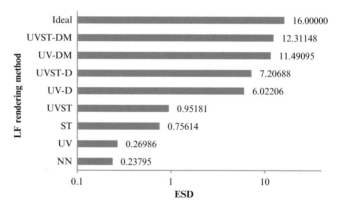

Fig. 2.11 Theoretical $\overline{\text{ESD}}$ for different rendering methods for $k = 0.4$m, $l = 0.05$cm, $\bar{d} = 10$m, and $\overline{\Delta d} = 1$m

2.5.4.2 Simulation Results

Figure 2.10 shows the simulated results, where the object depth d is changed from 10 m to 50 m with steps of 5 m to analyse the effect of d on rendering output distortion in PSNR for different rendering methods. The acquisition parameters are $k = 0.4$ m and $l = 0.05$ cm (i.e. camera resolution of 1024×768). Notice that all the parameters for camera configuration and rendering algorithm were set the same as those used to obtain the theoretical results shown in Fig. 2.9. 10 % depth error was added in the experiments. Figure 2.10 shows the average results calculated from 288,000 experiments for 9 depths, 8 rendering methods, 4 3D models, and 1000 virtual viewpoints for each experiment. As it can be seen, rendering

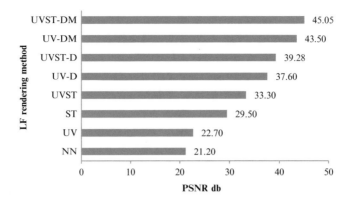

Fig. 2.12 Experimental rendering quality in PSNR for different LR methods

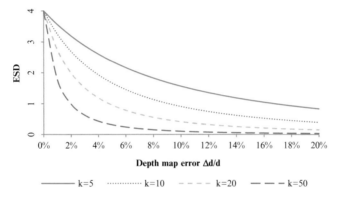

Fig. 2.13 Theoretical ESD for UV-DM for $\overline{d} = 100$, $\overline{\Delta d}$ in the range of $[0\%, 20\%]$, $l = 0.01$, $|\omega| = 4$ for $k = 5, 10, 20, 50$

methods with full depth information UVST-DM and then UV-DM performed the best with the least distortion (in PSNR) followed by rendering methods with focusing depth information of UVST-D and then UV-D. Not surprisingly, the blind rendering methods with no depth information had the highest distortion with UVST performing the best among blind methods followed by ST, UV, and NN. The distance of the scene to the camera grid had a direct effect on output distortion, where further distance caused higher distortion for all methods, more significantly for methods with depth information and less pronounced for blind methods. More importantly, the results show the same trends with the theoretical ESD values shown in Fig. 2.9.

Figure 2.12 shows the average PSNR values over 32,000 simulations at $d = 10$ m. NN interpolation performs the worst; UVST-DM is the best while UVST is the best blind rendering method. This order is consistent with the theoretically calculated ESD shown in Fig. 2.11.

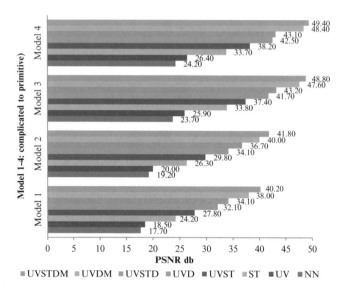

Fig. 2.14 Rendering quality and scene complexity

Figure 2.14 shows the mean PSNR from 144,000 experiments for different rendering methods, categorised based on the complexity of the scene. As can be seen, more complex scenes result in reduced rendering quality. This can be explained due to fixed ESD for different scenes with different complexities in terms of higher spatial frequency components. Nevertheless, ESD provides the right ranking on the performance amongst the various methods.

Figure 2.15 shows the rendering distortion from 144,000 experiments based on the distance of the virtual camera to the scene. As it is shown, far navigation results in higher rendering quality compared with closer observations. Again, this can be explained as a consequence of reduction in the required high-frequency components to be sampled. Note that this experiment is different from experiments demonstrated in Fig. 2.10 and that is why the results are different. In this experiment, the light-field system was fixed and the depth of virtual cameras was changed. In the previous experiment, the object depth is changed and the PSNR is calculated as the mean of 1000 random virtual cameras.

2.5.5   Results on Acquisition Configurations

By changing l and k, respectively, various LF acquisition configurations were simulated.

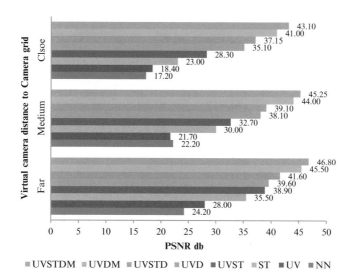

Fig. 2.15 Rendering quality and observation distance

2.5.5.1 Theoretical Expectation

Figure 2.16 demonstrates the theoretical relationship between k, the distance between the cameras in the camera grid, and ESD. As expected, for all methods, dense camera grid (small k) results in high ESD and therefore high rendering quality. In this figure, $d = 50$ m, $l = 0.05$ cm (camera resolution of 1024×768), and $k \in [0.1$ m$, 0.9$ m$]$ with the same assumption for depth error as the case shown in Fig. 2.9.

As it can be seen, changing the value of k has limited effects on UV-D/UVST-D and UV-DM/UVST-DM, though at large k, UV-D and UV-DM performance gets worse compared to UVST-D and UVST-DM, respectively. Also ESD of the ideal case (when there is no error in depth) is independent of k as demonstrated before. However, for blind methods, k has a significant effect on ESD values. NN, UV, ST, and UVST all perform poorly especially for a large k. This confirms the view that by utilising depth information, the cost of acquisition system can be significantly reduced.

Figure 2.18 presents the theoretical relationship between l, the pixel size, and ESD. It is clear that for all methods, high resolution (small l) results in high ESD and therefore high rendering quality. In this figure, $d = 50$ m, $k = 0.4$ m, and $l \in [0.02$ cm$, 0.1$ cm$]$, i.e. camera resolution of 2560×1920 to 512×384, respectively, with the same assumption for depth error as the case shown in Fig. 2.9.

As it can be seen, changing l has a direct effect on all methods. This effect is much more significant for UV-D, UVST-D, UV-DM, UVST-DM, and the ideal case and less significant for blind methods. NN/UV and also ST/UVST performed similarly especially for a small l (high resolution).

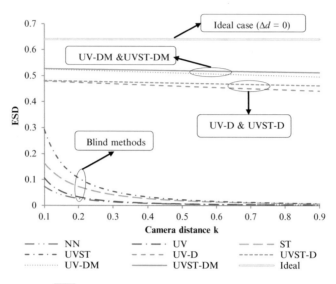

Fig. 2.16 Theoretical $\overline{ESD}$ for different LF rendering methods based on camera distance k between 0.1m and 0.9m for $l = 0.05$cm

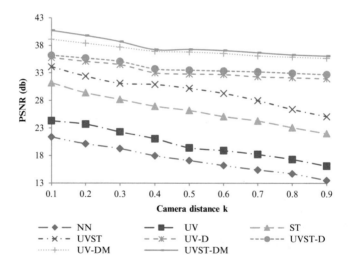

Fig. 2.17 Experimental rendering quality in PSNR for different LF rendering methods vs. camera distance k

2.5.5.2 Simulation Results

Experiments were carried out to see the effect of k in rendering distortion in terms of PSNR so as to make a comparison to the theoretical ESD values. In the first experiment, $d = 50$ m, object length $= 5$ m, $l = 0.05$ cm, and $k \in [0.1\text{ m}, 0.9\text{ m}]$ and 10 % depth error was added. Figure 2.17 shows the results calculated from

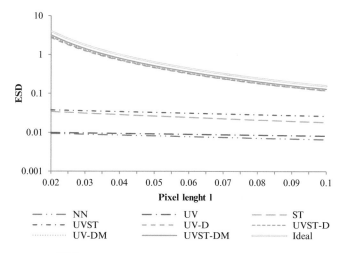

Fig. 2.18 Theoretical $\overline{\mathrm{ESD}}$ for different LF rendering methods based on pixel length l between 0.02cm (camera resolution of 2560×1920) and 0.1cm (camera resolution of 512×384)

random 288,000 trials. As it can be seen, large separation between the cameras decreases the rendering PSNR as expected. However, the impact of increasing k is less significant for UV-D, UVST-D, UV-DM, and UVST-DM compared to the blind methods.

The second experiment shows the relationship between the resolution of cameras (in terms of pixel length l) and the rendering distortion in terms of PSNR. In this experiment $d = 50$ m, object length $= 5$ m, $k = 0.4$ m, and $l \in [0.02$ cm, 0.1 cm], i.e. resolution of 2560×1920 to 512×384, respectively, and 10 % depth error. Figure 2.19 illustrates the results calculated from 288,000 trials. As it can be seen, high resolution (smaller value of l) increases the rendering PSNR as expected. However, l has less impact on the blind rendering methods and more on UV-D, UVST-D, UV-DM, and UVST-DM.

Therefore, the theoretical expectations based on ESD analysis are confirmed by the empirical results. This can be seen clearly by comparing Fig. 2.16 with Fig. 2.17 and Fig. 2.18 with Fig. 2.19. Notice that the theoretical expectation is shown in ESD while the simulation results are shown in PSNR, and their relationship will be examined in the next section.

2.5.6 Discussions

Figures 2.9, 2.10, 2.11, 2.12, 2.13, 2.14, 2.15, 2.16, 2.17, 2.18, and 2.19 present the theoretical expectations in terms of ESD and experimental results in terms of PSNR for different scenarios. To verify whether ESD is a good distortion indicator,

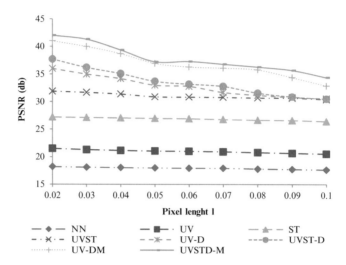

Fig. 2.19 Experimental rendering quality in PSNR for different LF rendering methods vs. pixel length *l*

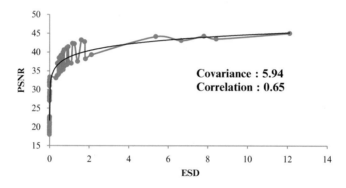

Fig. 2.20 Theoretical calculated ESD from Fig. 2.9 vs. experimental PSNR from Fig. 2.10, both obtained by changing the object depth ($\bar{d}$ from 10 to 50 m)

an analysis was conducted of ESD vs. its counterpart PSNR, i.e. pairs of Figs. (2.9, 2.10), (2.16, 2.17) and (2.18, 2.19).

Figure 2.20 shows the average experimental PSNR from Fig. 2.10 vs. theoretical ESD from Fig. 2.9, both obtained by changing the object depth $\bar{d}$. The trendline, covariance, and correlation of PSNR vs. ESD are also shown in Fig. 2.20.

Similarly, Fig. 2.21 demonstrates the observed PSNR from Fig. 2.17 vs. calculated ESD from Fig. 2.16, both obtained by changing the camera density. Again, the trendline, covariance, and correlation of PSNR vs. ESD are shown.

Figure 2.22 shows the observed PSNR from Fig. 2.19 vs. calculated ESD from Fig. 2.18, both obtained by changing the camera resolution.

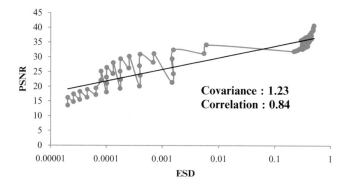

Fig. 2.21 Theoretical calculated ESD from Fig. 2.16 vs. experimental PSNR from Fig. 2.17, both obtained by changing the camera density (*k* from 1 to 9 m)

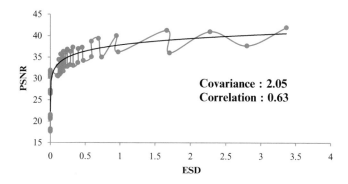

Fig. 2.22 Theoretically calculated ESD from Fig. 2.18 vs. experimental PSNR from Fig. 2.19, both obtained by changing the resolution (*l* from 0.02 to 0.1 cm)

Figures 2.20, 2.21, and 2.22 show a high correlation between theoretically calculated ESD and observed PSNR. In addition, as the trendlines demonstrate, there is an empirical relationship that can be explored to estimate output distortion in PSNR directly from calculated ESD without experiments. This will be explored in the next section.

2.6   Empirical Relationship Between ESD and PSNR

The experiments have shown that there is a relationship between ESD and PSNR. Since PSNR is a function of MSE (mean squared error), it is expected that MSE is a function of $\overline{\text{ESD}}$ for each given LF rendering method, denoted by $\text{ESD}_{\text{method}}$, and for a given fixed scene, i.e. $\text{MSE} = f(\text{ESD}_{\text{method}})$. In general, empirical f can be formulated as

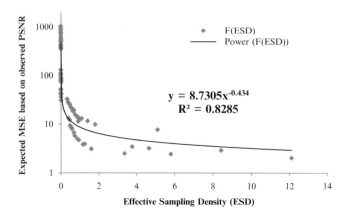

Fig. 2.23 A general curve fitting for $f(ESD)$ estimation based on calculated $\overline{ESD}$ vs. expected MSE

$$f(\mathrm{ESD_{method}}) = Q \times \mathrm{ESD_{method}}^P \qquad (2.12)$$

To find f, a subset of existing data is chosen as training set for curve fitting and the rest of the data as a validation set to test the accuracy of the empirical model f. To generate the curve fitting data, a map between observed PSNR and expected MSE is calculated as follows:

$$f(\mathrm{ESD_{method}}) = \mathrm{Expected\ MSE} = \frac{255^2}{10^{\left(\frac{\mathrm{Observed\ PSNR}}{10}\right)}} \qquad (2.13)$$

The data presented in Figs. 2.9 and 2.10 (theoretical and experimental results based on changing the object depth) is used as the training set and data demonstrated in Figs. 2.16, 2.17, and Figs. 2.18, 2.19 for validation. Figure 2.23 demonstrates the overall curve fitting. This curve fitting is done on all the data and without clustering the data based on the rendering methods. Figure 2.24 shows the curve fitting for each LF rendering method separately (method dependent). The optimum value for $f(\mathrm{ESD_{method}})$ for best estimation is when it is equal to expected MSE.

Figure 2.25 shows a summary of curve fitting and validation errors of PSNR estimation for all LF rendering methods. As it can be seen from Fig. 2.25, the method-dependent estimation error for validation tests is less than 3 %. If the method-dependent equations are not available, the estimation error for the overall equation is less than 12 %. This shows that empirical equations for $f(\mathrm{ESD_{method}})$ are accurate to indicate the rendering distortion in terms of PSNR. These equations offer a way to directly estimate the overall rendering distortion of an LF-based FVV system from the calculated ESD without implementation and experiments.

By applying the analytical ESD equations to the proposed empirical equations, a direct model to estimate the rendering quality in PSNR from LF system parameters can be formulated. This helps the system designers to optimise the LF acquisition

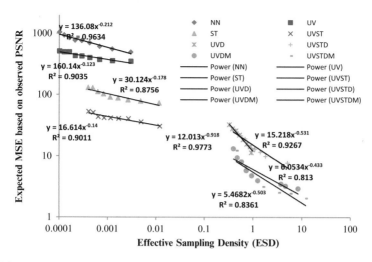

Fig. 2.24 Method-dependent curve fittings for $f(\text{ESD}_{\text{method}})$

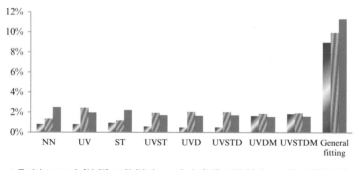

🔲 Training set : d∈[10,50]   🔳 Validation set I : k∈[1,9]   ■ Validation set II : l∈[0.02,0.1]

Fig. 2.25 Summary of curve fitting training and validation errors of PSNR estimation

and LF rendering components without exhaustive experimental implementation of each configuration. For instance, for a general UVDM($d, \Delta d, k, l, |\omega|$) method, by applying the ESD from Eq. (2.11), the rendering distortion can be directly calculated as

$$\text{PSNR}_{\text{UVDM}(d,\Delta d,k,l,|\omega|)} \cong 20\log_{10}\frac{255}{\sqrt{3.4545\left(\frac{|\omega|}{\left[l(d+\Delta d)+\frac{\Delta d \times k}{d}\left(\sqrt{|\omega|}-1\right)\right]^2}\right)^{-0.256}}} \tag{2.14}$$

Table 2.5 summarises the empirical boundaries of Q and P for different LF rendering methods, estimated for different scenes and acquisitions.

Table 2.5 Empirical boundaries of P and Q

LF rendering method type	LF rendering method	Q	P
LF rendering methods with no depth information $10 < Q < 300$ $-0.3 < P < -0.1$	NN	$50 < Q_{NN} < 300$	$-0.3 < P_{NN} < -0.2$
	ST	$20 < Q_{ST} < 200$	$-0.2 < P_{ST} < -0.1$
	UV	$20 < Q_{UV} < 250$	$-0.25 < P_{UV} < -0.1$
	UVST	$10 < Q_{UVST} < 200$	$-0.2 < P_{UVST} < -0.1$
LF rendering methods with focusing depth information $10 < Q < 40$ $-1.0 < P < -0.15$	UVD	$10 < Q_{UVD} < 40$	$-1.0 < P_{UVD} < -0.15$
	UVSTD	$10 < Q_{UVSTD} < 40$	$-1.0 < P_{UVSTD} < -0.15$
LF rendering methods with full depth information $1 < Q < 15$ $-0.9 < P < -0.2$	UVDM	$1 < Q_{UVDM} < 15$	$-0.9 < P_{UVDM} < -0.2$
	UVSTDM	$1 < Q_{UVSTDM} < 15$	$-0.9 < P_{UVSTDM} < -0.2$
	General method	$1 < Q < 10$	$-1.4 < P < -0.2$

The differences in $f(\text{ESD}_{\text{method}})$ equations can be directly explained due to differences in the scene complexities and interpolation methods. Despite these differences, the general model offers a good indication on what the overall distortion in terms of PSNR should be expected by a given $\overline{\text{ESD}}$.

2.7 Subjective Assessment

While previous section discussed the correlation between ESD and output video distortion in terms of PSNR, this section demonstrates that ESD is also highly correlated with subjective assessment of the perceived video quality. A subjective quality assessment based on ITU-T standardisation and guidelines on "subjective video quality assessment methods for multimedia applications" [25] and using degradation category rating (DCR) method was carried out. The test procedure is based on recommendations proposed in VQEG reports [59, 60]. Three rendering methods, UVST as a candidate of rendering methods with no depth information, UV-D with focusing depth, and UV-DM with full depth information, were selected for subjective test. The ground truth from the simulator and Stanford light-field archive [61] was used as reference images. The original Stanford camera grid to capture real scenes is 17×17, i.e. 289 reference images. To provide the ground truth for real scenes with real depth values, a subset of these reference images as a sparse 8×8 camera grid was selected for acquisition component and a subset of other cameras were used as ground truth. Eighteen subjects participated in the test. For each of the three candidate rendering methods, eight rendering outputs from different viewpoints for four different scenes, "*chess board*" and "*room*" from simulator and "*eucalyptus flowers*" and "*Lego knights*" from Stanford real data, were generated. These 96 test sequences as a pair of reference and rendering output were presented to each subject with the recommended time pattern and experiment conditions as proposed in [25, 62]. The subjects were asked to rate the impairment of the second stimulus in relation to the reference into one of the five-level scales: 5—Imperceptible, 4—Perceptible but not annoying, 3—Slightly annoying, 2—Annoying, and 1—Very annoying.

The ESD is also calculated for each pair of scene and rendering method using the equations presented in Tables 2.1 and 2.3. There are totally 12 values for ESD (4 scenes and 3 rendering methods). Each value of ESD is corresponded to eight different views.

Figure 2.26 shows samples of the test sequences, presented to the subject panel. Note that Fig. 2.26 shows 12 different pairs out of 96 test sequences which were presented to each subject.

Figure 2.27 illustrates the results of the subjective test for each rendering method. The average and variance of the impairment for each rendering method were calculated from 576 collected scores (32 test sequences among 18 subjects).

To validate the relationship between ESD and subjective DCR rating, the procedure for specifying accuracy and cross-calibration of video quality metrics

Fig. 2.26 Samples of test sequences used in the subjective assessment

Fig. 2.27 Subjective
assessment of three LF
rendering methods by using
degradation category rating
(DCR), showing the mean
and variance of rating from
576 collected scores for each
method (32 test sequences
among 18 subjects) with a
five-level scale for rating the
impairment

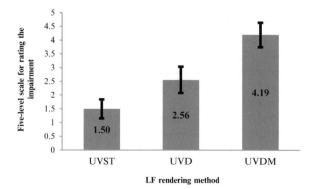

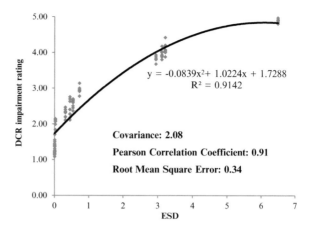

Fig. 2.28 DCR impairment rating for subjective assessment vs. theoretical ESD and the empirical relationship between these two parameters

$y = -0.0839x^2 + 1.0224x + 1.7288$
$R^2 = 0.9142$

Covariance: 2.08

Pearson Correlation Coefficient: 0.91

Root Mean Square Error: 0.34

proposed in VQEG reports [59, 60] was employed. Figure 2.28 shows the scatter plot for the ESD-DCR couples for all 96 test sequences. Please note that for each eight test sequences for different views, there is only one calculated ESD. To obtain the empirical relationship between DCR impairment rating and ESD, a polynomial curve fitting, as one of the candidates in VQEG reports, is applied over the data. The *Pearson correlation coefficient* is calculated as 0.91 which demonstrates a high relationship among ESD and DCR. The curve fitting has a *root mean square error* of 0.34 which shows around 10 % error to predict DCR from calculated ESD which is technically satisfactory.

Figure 2.29 shows an outdoor scene rendered with the proposed FVV system for subjective comparison of ground truth with the rendered output.

2.8   Application of ESD

2.8.1   Calculating the Minimum Number of Cameras

Regular camera grids are widely used for FVV acquisition. Several studies are reported to calculate the minimum number of cameras for regular grids which can be categorised into three main approaches: (a) plenoptic signal spectral analysis [3, 24] and the light-field spectral and frequency analysis [4, 5], (b) view interpolation geometric analysis such as [6], and (c) optical analysis of light field [14, 35, 36]. However, these methods are essentially based on several simplifying assumptions (e.g. Lambertian scene, no occlusion, linear interpolation over 4 or 16 rays, and calculating the Nyquist sampling rate without considering under-sampling), and also suggest an impractically high number of cameras [28, 29].

In contrast, using ESD to address this problem has several advantages such as studying under-sampled light field under realistic conditions (non-Lambertian

Fig. 2.29 An outdoor scene, ground truth, and the rendered output for subjective comparison

Ground Truth

Rendered Output

reflections and occlusions) and rendering with complex interpolations. The optimisation method based on ESD proposed in [28, 29] is summarised here.

In $\mathrm{ESD}_{\mathrm{UVDM}(d,\Delta d,k,l,|\omega|)}$ expression given as Eq. (2.11), d is given by scene geometry and Δd is determined by the depth estimation method and cannot be altered by us. Changing the other three parameters could potentially improve the rendering quality. By assuming a given camera resolution, i.e. a fixed value of l, two other parameters can be tuned to compensate for the depth estimation error while maintaining the rendering quality. These parameters include k as a measure of density of cameras during acquisition and $|\omega|$ as an indicator of complexity of rendering method. ESD is proportional to $|\omega|$ and inversely proportional to k, i.e. higher camera density (smaller k) and employing more rays for interpolation results in higher ESD. The optimisation of k is summarised here and optimisation of $|\omega|$ will be discussed in next subsection.

The problem of calculating the minimum number of cameras can be expressed in terms of minimum camera density, i.e. maximum k to provide required ESD in each point of the scene to compensate for the adverse effect of depth map estimation errors. This minimum required ESD can be calculated for the ideal case when there is no error in depth estimation and there are n rays employed for interpolation. Hence the optimisation method can be written as follows:

Find the maximum k to satisfy

$$\text{ESD}_{\text{UVDM}(d,\Delta d,k,l,|\omega|)} = \text{ESD}_{\text{Ideal}} \rightarrow \text{ESD}_{\text{UVDM}(d,\Delta d,k,l,|\omega|)} = \text{ESD}_{\text{UVDM}(d,0,k,l,n)} \rightarrow$$

$$k = \frac{ld\left(d\sqrt{\frac{|\omega|}{n}} - d - \Delta d\right)}{\Delta d\left(\sqrt{|\omega|} - 1\right)} = \frac{l\left(\left(\sqrt{\frac{|\omega|}{n}} - 1\right)d^2 - d\Delta d\right)}{\Delta d\left(\sqrt{|\omega|} - 1\right)}$$

(2.15)

where

$$\Delta d > 0 \text{ and } |\omega| > n\left(\frac{d + \Delta d}{d}\right)^2$$

Figure 2.30 shows the summary of theoretical expectations and experimental results for the optimisation process. Figure 2.30a, b illustrates the theoretical expectations. It is assumed that $l = 0.01$, average depth of scene $\bar{d} = 100$, relative depth map error $\frac{\Delta d}{d}$ between 1 % and 20 %, and $|\omega|$ is calculated as follows to satisfy the condition of Eq. (2.15): $|\omega| > 4\left(\frac{100+20}{100}\right)^2 > 5.76 \rightarrow |\omega| = 6$. For any given depth estimation error $\Delta d \leq 20\%$, k is calculated directly from Eq. (2.15) to maintain $\overline{\text{ESD}}$ at 4.00, the ideal ESD calculated for $n = 4$ and $\Delta d = 0$. Figure 2.30a demonstrates the ESD for fixed $k = 14.4$ and optimum k calculated from Eq. (2.15). Figure 2.30b shows the calculated k in such a scenario. The corresponding point for 10 % error in depth estimation is highlighted in Fig. 2.30a, b, respectively, to show the relation of these two figures. Figure 2.30c shows that the rendering PSNR is maintained at a prescribed value (for instance 50 dB) with calculated k in contrast with the average

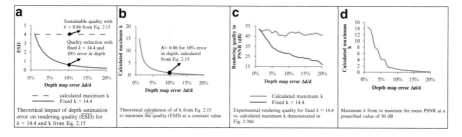

Fig. 2.30 Summary of theoretical and experimental optimisation of k (camera density) based on ESD

PSNR for fixed $k = 14.4$; the required k to maintain the quality is demonstrated in Fig. 2.30d. Figure 2.30 shows that for high error rates, changing k using Eq. (2.15) results in significant improvements over the fixed camera density and can maintain the quality around the prescribed 50 dB.

2.8.2 Calculating the Minimum Interpolation Complexity

The number of rays selected by *ray selection* process of a given rendering method is an important parameter of the rendering complexity. On the one hand, increasing the number of rays results in increasing ESD in each point of the scene resulting in higher output quality. On the other hand, this also increases the interpolation complexity resulting in slower rendering which might not be acceptable in real-time applications. To calculate the optimum number of rays for interpolation to satisfy both required rendering quality and rendering efficiency, an optimisation method is proposed in [28, 31].

With the same approach as in previous subsection the minimum $|\omega|$ to avoid quality deterioration due to errors in depth maps can be calculated as

Find the minimum $|\omega|$ to satisfy

$$\text{ESD}_{\text{UVDM}(d,\Delta d,k,l,|\omega|)} = \text{ESD}_{\text{Ideal}} \rightarrow \text{ESD}_{\text{UVDM}(d,\Delta d,k,l,|\omega|)}$$

$$= \text{ESD}_{\text{UVDM}(d,0,k,l,n)} \rightarrow |\omega| = \left(\frac{l(d + \Delta d) - \frac{\Delta d \times k}{d}}{\frac{ld}{\sqrt{n}} - \frac{\Delta d \times k}{d}} \right)^2 \quad (2.16)$$

where

$$k < \frac{ld^2}{\Delta d \sqrt{n}}$$

Figure 2.31 shows the summary of theoretical expectations and experimental results for the optimisation process.

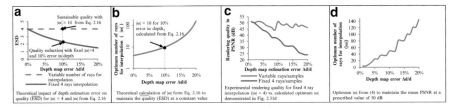

Fig. 2.31 Summary of theoretical and experimental optimisation of $|\omega|$ (number of rays employed in interpolation)

Figure 2.31a, b shows the theoretical expectations for this optimisation model. l, $\overline{d}$, and $\overline{\Delta d}$ are the same as in Fig. 2.30. k is calculated as follows to satisfy the condition of Eq. (2.16): $k < \frac{0.01 \times 100^2}{20\sqrt{4}} < 2.5 \rightarrow k = 2.2$. For any $\Delta d < 20\%$, $|\omega|$ is calculated directly from Eq. (2.16) to maintain $\overline{ESD}$ at 4.00, the ideal ESD calculated for $n = 4$. Figure 2.31a demonstrates the ESD for fixed four-ray interpolation and for optimum number of rays calculated from Eq. (2.16). Figure 2.31b shows the actual number of rays $|\omega|$, employed in interpolation in such a scenario. The corresponding point for 10 % error in depth estimation is highlighted in Fig. 2.31a, b, respectively, to show the relation of these two figures. Figure 2.31c shows that the rendering PSNR is maintained at a prescribed value (for instance 50 dB) with calculated optimum number of rays $|\omega|$ in contrast with the average PSNR for conventional fixed four-ray interpolation, and calculated number of rays $|\omega|$ is demonstrated in Fig. 2.31d. Figure 2.31 shows that for high level of error in depth, the use of optimum $|\omega|$ using Eq. (2.16) results in significant improvements over the conventional fixed four-ray interpolation and can maintain the rendering quality around the prescribed 50 dB.

2.8.3 Irregular Acquisition Based on the Scene Complexity

As noted before, FVV acquisition is typically performed by using a regular camera grid. While a regular acquisition itself results in non-uniform sampling density, this non-uniformity does not match the scene complexity and frequency variations. The simplest non-uniform acquisition can be done by using an irregular camera grid. The problem is then to find the positions and orientations of the camera in the gird to provide higher ESD in the parts of the scene with higher complexity and vice versa. The theory of irregular/non-uniform signal sampling has been widely investigated and it is shown that irregular sampling can reduce the number of required samples for perfect reconstruction of the signal. However to the best of our knowledge, this property has not been explored for FVV acquisition and rendering. An optimisation method based on ESD for this problem is proposed in [28, 30]. It is shown that ESD can be regarded as a set of utility functions $U_h(ESD)$ based on the given scene complexity factor h. The higher the scene complexity, more ESD would be required for a given reconstruction fidelity. Each acquisition configuration and rendering method result in an ESD pattern, which varies in the scene space. Assume that the scene could be partitioned into a number of smaller 3D regions or blocks, each having a fixed average complexity h, determined from the highest frequency components of the block computed by applying DCT transform. Then, the aim of the optimisation problem could be to find the optimum acquisition configuration which provides the minimum required ESD for all blocks. This optimisation problem is discussed in [28, 30] and is shown that an analytical dynamic programming solution is available to compute the optimum irregular camera grid.

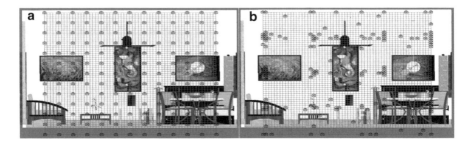

Fig. 2.32 (**a**) Regular camera grid with 169 (13 × 13) cameras; (**b**) optimum irregular camera grid for 169 cameras

Theoretical analysis and experimental validation showed that the output video quality can be significantly improved (around 20 % in mean PSNR) by employing the proposed irregular acquisition compared with the regular camera grid. Figure 2.32 shows the initial regular camera grid and the optimum irregular camera grid for 169 cameras. The average of rendering PSNR from 1000 virtual cameras was improved from 39.10 dB for regular grid to 46.60 dB for optimum irregular grid.

2.9 Conclusion

This chapter has discussed the concept of ESD and its application in FVV quality assessment, and comparison, evaluation, and optimisation of FVV acquisition and rendering subsystems. Using ESD, different LF rendering methods and LF acquisition configurations can be theoretically evaluated and compared. Eight well-known rendering methods with different acquisition configurations have been analysed through ESD and simulation. The results have shown that ESD is an effective indicator of distortion that can be obtained directly from system parameters and takes into consideration both acquisition and rendering. In addition, an empirical relationship between the theoretical ESD and achievable PSNR has been established. Furthermore, a subjective assessment has confirmed that ESD is highly correlated with the perceived output quality. Finally several problems on FVV evaluation and optimisation have been approached by using ESD. This has been done by analysing the impact of depth estimation errors on ESD and optimisation of ESD with respect to the *camera density* and *ray selection complexity* for a given output quality. Although this chapter focuses on the overall distortion of an LF-based FVV system, the concept is readily extended to measure the rendering quality at a specific location or part of the scene.

2.10   Biography

Hooman Shidanshidi graduated from the Bahá'í Institute for Higher Education (BIHE) University, Iran, with the degree of Bachelor of Software Engineering and received his Master of Research and Ph.D. in Computer Engineering from the University of Wollongong, Australia. He has been a Lecturer and Faculty Member at Bahá'í Institute for Higher Education (BIHE) University since 1998 and a Postdoctoral Research Fellow in ICT Research Institute at the University of Wollongong since 2013. Before joining the University of Wollongong, he was also the Senior Project Manager in several software development companies. His research areas include computer vision, multimedia signal processing, free viewpoint video, computational intelligence, and simulation optimisation.

Farzad Safaei graduated from the University of Western Australia with the degree of Bachelor of Engineering (Electronics) and obtained his Ph.D. in Telecommunications Engineering from Monash University, Australia. Currently, he is the Professor of Telecommunications Engineering and Managing Director of ICT Research Institute at the University of Wollongong. Before joining the University of Wollongong, he was the Manager of Internetworking Architecture and Services Section in Telstra Research Laboratories, Melbourne, Australia. His research interests include immersive multimedia communications and free viewpoint TV.

Wanqing Li received his Ph.D. in electronic engineering from The University of Western Australia. He joined Motorola Lab in Sydney (98-03) as a Senior Researcher and later a Principal Researcher and was a visiting researcher at Microsoft Research, Redmond, USA, in 2008, 2010, and 2013. He is currently an Associate Professor and Co-Director of Advanced Multimedia Research Lab (AMRL) of University of Wollongong, Australia. His research areas are 3D computer vision and 3D multimedia signal processing, including 3D reconstruction, human motion analysis, detection of objects and events, and free viewpoint video. Dr. Li is currently a co-chair of the 3D Rendering, Processing and Communications interest group, Multimedia Technical Committee of IEEE Communication Society. He is the guest editor of the special issue on human activity understanding from 2D and 3D data (2015), International Journal of Computer Vision, and the special issue on Visual Understanding and Applications with RGB-D Cameras (2013), Journal of Visual Communication and Image Representation. He served as a co-organizer of many IEEE international conferences and workshops.

References

1. Tanimoto M, Tehrani MP, Fujii T, Yendo T (2011) Free-viewpoint TV. IEEE Signal Process Mag 28:67–76
2. Tanimoto M (2012) FTV: free-viewpoint television. Signal Process Image Comm 27:555–570
3. Chai JX, Tong X, Chan SC, Shum HY (2000) Plenoptic sampling. Proc SIGGRAPH (ACM Trans Graphics) 307–318
4. Zhang C, Chen T (2003) Spectral analysis for sampling image-based rendering data. IEEE Trans Circ Syst Video Technol 13:1038–1050
5. Zhang C, Chen T (2006) Light field sampling. Synth Lect Image Video Multimed Process 2:1–102
6. Zhouchen L, Heung-Yeung S (2004) A geometric analysis of light field rendering. Int J Comput Vision 58:121–138
7. King-To N, Zhen-Yu Z, Chong W, Shing-Chow C, Heung-Yeung S (2012) A multi-camera approach to image-based rendering and 3-D/multiview display of ancient Chinese artifacts. IEEE Trans Multimed 14:1631–1641

8. Safaei F, Mokhtarian P, Shidanshidi H, Li W, Namazi-Rad M, Mousavinia A (2013) Scene-adaptive configuration of two cameras using the correspondence field function. In: IEEE international conference on multimedia and expo (ICME). pp 1–6
9. Takahashi K, Naemura T (2006) Layered light-field rendering with focus measurement. Signal Process Image Comm 21:519–530
10. Daniel NW, Daniel IA, Ken A, Brian C, Tom D, David HS et al (2000) Surface light fields for 3D photography. In: 27th annual conference on computer graphics and interactive techniques
11. Jingyi Y, McMillan L, Gortler S (2002) Scam light field rendering. In: 10th pacific conference on computer graphics and applications. pp 137–144
12. Shum HY, Sun J, Yamazaki S, Lin Y, Tang CK (2004) Pop-up light field: an interactive image-based modeling and rendering system. ACM Trans Graphics 23:143–162
13. Wen W, Jiang Zhang Z, Yao Si C, Zeng D (2010) An efficient method for all-in-focused light field rendering. In: 3rd IEEE international conference on computer science and information technology (ICCSIT). pp 399–404
14. Aaron I, Leonard M, Steven JG (2000) Dynamically reparameterized light fields. In: 27th annual conference on computer graphics and interactive techniques
15. Hansung K, Guillemaut JY, Takai T, Sarim M, Hilton A (2012) Outdoor dynamic 3-D scene reconstruction. IEEE Trans Circ Syst Video Technol 22:1611–1622
16. Liu SX, An P, Zhang ZY, Zhang Q, Shen LQ, Jiang GY (2009) High quality virtual view synthesis based on corrected surface mapping and image fusion. Electron Lett 45:30–32
17. Ekmekcioglu E, Velisavljevic XV, Worrall ST (2011) Content adaptive enhancement of multi-view depth maps for free viewpoint video. IEEE J Selected Topics Signal Process 5:352–361
18. Scandarolli T, de Queiroz RL, Florencio DA (2013) Attention-weighted rate allocation in free-viewpoint television. IEEE Signal Process Lett 20:359–362
19. Qifei W, Xiangyang J, Qionghai D, Naiyao Z (2012) Free viewpoint video coding with rate-distortion analysis. IEEE Trans Circ Syst Video Technol 22:875–889
20. Zhun H, Qionghai D (2007) A new scalable free viewpoint video streaming system over IP network. In: IEEE international conference on acoustics, speech and signal processing (ICASSP). pp II-773-II-776
21. Adelson EH, Bergen JR (1991) The plenoptic function and the elements of early vision. In: Computational models of visual processing. Vision and Modeling Group, Media Laboratory, Massachusetts Institute of Technology, pp 3–20
22. Levoy M, Hanrahan P (1996) Light field rendering. Proc SIGGRAPH (ACM Trans Graphics) 31–42
23. Gortler SJ, Grzeszczuk R, Szeliski R, Cohen MF (1996) The lumigraph. Proc SIGGRAPH (ACM Trans Graphics) 43–54
24. Do MN, Marchand-Maillet D, Vetterli M (2012) On the bandwidth of the plenoptic function. IEEE Trans Image Process 21:708–717
25. ITU-T Recommendation P (1999) Subjective video quality assessment methods for multimedia applications
26. Shidanshidi H, Safaei F, Li W (2011) Objective evaluation of light field rendering methods using effective sampling density. In: IEEE international workshop on multimedia signal processing (MMSP). pp 1–6
27. Shidanshidi H, Safaei F, Li W (2015) Estimation of signal distortion using effective sampling density for light field based free viewpoint video. IEEE Trans Multimed 17(10):1677–1693
28. Shidanshidi H (2014) Effective sampling density for quality assessment and optimization of light field rendering and acquisition. Doctor of Philosophy Thesis, School of Electrical, Computer and Telecommunications Engineering, University of Wollongong
29. Shidanshidi H, Safaei F, Li W (2013) A method for calculating the minimum number of cameras in a light field based free viewpoint video system. In: IEEE international conference on multimedia and expo (ICME). pp 1–6
30. Shidanshidi H, Safaei F, Zamani-Farahani A, Li W (2013) Non-uniform sampling of plenoptic signal based on the scene complexity variations for a free viewpoint video system. In: IEEE international conference on image processing (ICIP). pp 3147–3151

31. Shidanshidi H, Safaei F, Li W (2015) Optimization of the number of rays in interpolation for light field based free viewpoint systems. In: IEEE international conference on multimedia and expo (ICME). pp 1–6
32. Shidanshidi H, Safaei F, Li W (2015) Effective sampling density and its applications to the evaluation and optimization of free viewpoint video systems. IEEE COMSOC MMTC E-Lett 10(2):21–25
33. Shidanshidi H, Safaei F, Li W (2016) Optimization of free viewpoint video acquisition and rendering subsystems by using effective sampling density. IEEE Trans Multimedia TBA
34. Camahort E, Lerios A, Fussell D (1998) Uniformly sampled light fields. Rendering Tech 98:117–130
35. Feng T, Shum HY (2000) An optical analysis of light field rendering. In: Fifth Asian conference on computer vision. pp 394–399
36. Lumsdaine A, Georgiev T (2008) Full resolution lightfield rendering. Indiana Univ Adobe Syst Tech Rep
37. Stewart J, Yu J, Gortler SJ, McMillan L (2003) A new reconstruction filter for undersampled light fields. In: 14th Eurographics workshop on rendering, Leuven, Belgium
38. Wenfeng L, Jin Z, Baoxin L, Sezan MI (2009) Virtual view specification and synthesis for free viewpoint television. IEEE Trans Circ Syst Video Technol 19:533–546
39. Zitnick CL, Kang SB, Uyttendaele M, Winder S, Szeliski R (2004) High-quality video view interpolation using a layered representation. Proc Siggraph (ACM Trans Graphics) 600–609
40. Seitz SM, Curless B, Diebel J, Scharstein D, Szeliski R (2006) A comparison and evaluation of multi-view stereo reconstruction algorithms. In: CVPR. pp 519–528
41. Kilner J, Starck J, Guillemaut JY, Hilton A (2009) Objective quality assessment in free-viewpoint video production. Image Commun 24:3–16
42. Sheikh HR, Bovik AC (2006) Image information and visual quality. IEEE Trans Image Process 15:430–444
43. Pons A, Malo J, Artigas J, Capilla P (1999) Image quality metric based on multidimensional contrast perception models. Displays 20:93–110
44. Winkler S (1998) A perceptual distortion metric for digital color images. In: ICIP, vol 3. pp 399–403
45. Brandão T, Queluz P (2006) Towards objective metrics for blind assessment of images quality. In: IEEE international conference on image processing (ICIP). pp 2933–2936
46. Seshadrinathan K, Bovik AC (2007) A structural similarity metric for video based on motion models. In: IEEE international conference on acoustics, speech, and signal processing, vol 1, pp I-869–I-872
47. Winkler S (2007) Video quality and beyond. In: European signal processing conference. pp 3–7
48. Wang Z, Bovik AC, Sheikh HR, Simoncelli EP (2004) Image quality assessment: from error visibility to structural similarity. IEEE Trans Image Process 13:600–612
49. Eskicioglu AM, Fisher PS (1995) Image quality measures and their performance. IEEE Trans Commun 43:2959–2965
50. Avcıbaş İ, Sankur B, Sayood K (2002) Statistical evaluation of image quality measures. J Electron Imag 11:206
51. Bosc E, Pepion R, Le Callet P, Koppel M, Ndjiki-Nya P, Pressigout M et al (2011) Towards a new quality metric for 3-D synthesized view assessment. IEEE J Selected Topics Signal Process 5:1332–1343
52. Bosc E, Koppel M, Pepion R, Pressigout M, Morin L, Ndjiki-Nya P et al (2011) Can 3D synthesized views be reliably assessed through usual subjective and objective evaluation protocols? In: 18th IEEE international conference on image processing (ICIP). pp 2597–2600
53. Raskar R, Agrawal AK (2010) 4D light field cameras. Google Patents (ed)
54. Takahashi K (2012) Theoretical analysis of view interpolation with inaccurate depth information. IEEE Trans Image Process 21:718–732

55. Shidanshidi H, Safaei F, Li W (2011) A quantitative approach for comparison and evaluation of light field rendering techniques. In: IEEE international conference on multimedia and expo (ICME). pp 1–4

56. Schwarz S, Olsson R, Sjostrom M (2013) Depth sensing for 3DTV: a survey. IEEE Multimed 20:10–17

57. Khoshelham K, Elberink SO (2012) Accuracy and resolution of kinect depth data for indoor mapping applications. Sensors 12:1437–1454

58. Pattinson T (2010) Quantification and description of distance measurement errors of a time-of-flight camera. M.Sc. Thesis, University of Stuttgart, Stuttgart, Germany

59. (2001) Methodological framework for specifying accuracy and cross-calibration of video quality metrics. Tech. Rep. T1.TR.72-2001

60. Brill MH, Lubin J, Costa P, Wolf S, Pearson J (2004) Accuracy and cross-calibration of video quality metrics: new methods from ATIS/T1A1. Signal Process Image Comm 19:101–107

61. The (new) Stanford light field archive. Stanford University Computer Graphics Laboratory, [Online]. http://lightfield.stanford.edu/lfs.html

62. Mantiuk RK, Tomaszewska A, Mantiuk R (2012) Comparison of four subjective methods for image quality assessment. In: Computer graphics forum, vol 31 (no. 8). Blackwell Publishing Ltd., pp 2478–2491

Chapter 3
Visual Quality-Regulated Three-Dimensional Video Coding (3-DVC)

Hong Ren Wu, Damian M. Tan, and David Wu

This chapter reviews (in Sect. 3.1) issues associated with research and development of three-dimensional (3-D) video coding in the context of the current push towards better viewing quality and experience of visual communications, broadcasting, and entertainment represented by ultra-high definition (UHD) television (TV), three-dimensional digital video (3-DV), multi-view video (MVV), and free viewpoint TV (FTV). It highlights (in Sect. 3.2) an agonizing impasse in a much-needed paradigm shift for video coding design from a bitrate-driven to a visual quality-driven design approach [23, 24, 64] based on Shannon's entropy [44] and rate-distortion (R-D) theories [4, 45], making true the viewing quality and experience promised by the aforementioned visual communication technologies. The key is to deliver with precision a designated visual picture quality discernible by human viewers for an intended application, maximizing intended visual experience at no more cost than theoretically and practically necessary by design rather than by chance or "the best effort." It examines (in Sect. 3.3) the state of the art in three key areas closely related to perceptual 3-D video coding, including 3-D video coding (3-DVC), perceptual video coding, and visual quality assessment and perceptual quality/distortion metric design. Issues associated with design of visual quality-regulated 3-DVC are discussed (in Sect. 3.4) for the transition from bitrate-driven design to visual quality-driven design, identifying a number of challenges for further research and investigations. It provides (in Sect. 3.5) insights into a comprehensive

H.R. Wu (✉)
Royal Melbourne Institute of Technology, Melbourne, VIC, Australia
e-mail: henry.wu@rmit.edu.au

D.M. Tan • D. Wu
HD² Technologies Pty. Ltd., Melbourne, VIC, Australia
e-mail: damian.tan@hd2tech.com; david.wu@hd2tech.com

© Springer Science+Business Media New York 2017
A. Kondoz, T. Dagiuklas (eds.), *Connected Media in the Future Internet Era*,
DOI 10.1007/978-1-4939-4026-4_3

mathematical formulation and illustration of a theoretical framework for perceptual quality-regulated encoding of 3-D video in terms of just noticeable differences (JNDs) and techniques applicable to various coding components of the framework.

3.1   Towards Better Viewing Quality and Experience

Technological advances in visual communications, broadcasting, and entertainment continue to captivate the general public, offering new heights of viewing quality and experience which aims at clarity, precision, and visual sensation by increasing spatiotemporal resolutions and bit depths of ultra-high definition TV/video [19], and realism and visual impact by three-dimensional digital video, i.e., 3-DV [54]. Multi-view video which goes beyond the stereoscopic 3-D video [6] and free viewpoint TV/video offer a much higher degree of freedom and choice of viewpoints or viewing perspectives [51].

A price has to be paid though, in order to deliver the quality increments and heightened visual experience promised by the aforementioned new video formats. For example, there is an eightfold increase in the uncompressed data rate for 3-D full high-definition (HD) video compared with the standard definition (SD) video [6, 54]. For an 8K (i.e., 7680×4320 pixels per frame) UHD video in its maximum configuration (12-bit coding of each color component in 4:4:4 component format with 120 frames per second progressive scan), there incurs a transmission data rate over 143 Gbps (gigabits per second) in uncompressed form [19] in contrast with approximately 166 Mbps (megabits per second) of uncompressed SD 4:2:2 component video [16]. Multi-view video beyond the stereoscopic 3-D rendition and free viewpoint TV yields even higher data rates and requires more advanced modelling and/or synthesis techniques [34, 51]. As a result, transportation of videos in these new formats puts an upward pressure on bandwidth demands of communication networks [1] and terrestrial broadcast bandwidth allocations [3, 34, 53].

Visual signal/data compression, propelled by digital computer technology with ever-increasing computational power and capabilities with reduced cost, was introduced, and has since been standardized with noticeable incremental improvements over the years and used to alleviate the pressure on transmission and storage demands, which has enabled now widespread visual communication, broadcasting, and entertainment applications and services [23, 64]. As elaborated previously, there still exists a significant discrepancy or gap between the bandwidth required by the aforementioned new video formats and applications and that is currently provided by communication networks at a reasonable and acceptable cost to populous consumers. To overcome this gap and to make the aforementioned applications and services economically viable, visual communication product manufactures and service providers have often adopted the use of high signal compression ratios based on a predominately rate-driven approach to visual signal compression, where visible and often uneven distortions, distortion variations, and uncontrolled distortions [71]

Fig. 3.1 Video coding artifacts. (**a**) An original frame from "Vidyo1"; (**b**) a frame difference related to (**a**) (with an offset of 128 used for display); (**c**) a frame difference corresponding to (**b**) from video coded by H.265/HEVC with quantization parameter QP = 42 (courtesy of K. Yang and S. Wan)

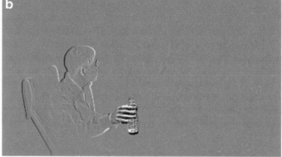

become commonplace and detrimental to quality and experience intended by digital services in various video formats. For example, Fig. 3.1 shows an original frame of test video sequence "Vidyo1," a difference frame generated by the original sequence, and its counterpart generated from reconstructed sequence encoded by an implementation of the latest H.265/HEVC standard [22, 37, 68]. Figure 3.1c reveals visible blocking, ringing, DCT basis image artifacts, and temporal fluctuations. In visual inspection of the encoded video sequence, uneven spread of the distortions and variations in type of distortions and degree of distortion visibility can be easily observed across spatial and temporal domains [68].

In 3-DV compression, the application of existing standardized video coding induces additional, often more severe, visual distortions and discomfort due to perceivable difference in matching areas of the two (left and right) views above

Fig. 3.2 Illustration of inter-view masking effect and depth perception loss in stereoscopic 3-D viewing due to coding distortions [64]. From left to right, *top row*: (**a**) an H.264/AVC-coded left view image at 10 Mbps (frame 2 of 3-D video sequence 28 of RMIT3DV database [7]), (**b**) the right view coded at 2 Mbps, (**c**) the right view coded at 512 kbps; *bottom row*: (**d**) a cropped portion of (**a**), (**e**) a cropped portion of (**b**), and (**f**) a cropped portion of (**c**). When (**a**) and (**b**) displayed for 3-D viewing, coding distortions in (**b**) were masked by (**a**), while (**a**) and (**c**) were displayed to the viewer, depth perception was lost due to severe distortions in (**c**). (Focus your eyes on the center of (**b**) or (**e**) image assuming that the point of picture in the real world by the side-by-side visual representation is beyond the screen/page to see stereo 3-D image/frame)

the visual threshold or the just-not-noticeable difference (JNND) level as a result of the compression, and visible disparity-compensated prediction error fluctuations in the enhancement (e.g., right) view which differ from those in the base (e.g., left) view [54, 64], and variation or even loss of depth perception caused by visual quality differences above a certain threshold between the two views, when visual quality is not or cannot be adequately regulated or controlled [64] (cf. Fig. 3.2).

To address these issues and technical challenges, it is beneficial to re-examine the design paradigms which have influenced and dominated video coding to date, to interrogate fundamental design principles to fully understand their guiding concepts and potentials, and to explain why delivery of user-centric visual communication and entertainment services with a quantum leap in quality of experience calls for new and effective compression theory, methods, and techniques, in order for information and communications technology (ICT) research community and associated industries to meet tremendous challenges and to take opportunities to bring about economic and social benefits to the global society.

3.2 Video Coding Design Paradigms

Effective, efficient, and affordable digital visual communications have been made possible by data compression techniques based on Shannon's entropy theory for information lossless compression [24, 44] and rate-distortion (R-D) theory for information lossy compression [4, 24, 45]. The latter has supported most of the practical applications to date. Entropy theory-based coding design is obviously signal quality driven which mandates perfect reconstruction of a given original signal. Based on the R-D theory, two distinct design approaches are possible, i.e., bitrate-controlled (or constant bitrate) coding and quality (or distortion)-controlled (or constant quality) coding. For a constant bitrate coder to lay a claim to its effectiveness in compression performance, it has to hold a fixed bitrate and then to maximize picture quality (or to minimize picture distortion), whereas for a constant quality coder to be effective, it must, first and foremost, be able to hold a given picture quality and, then, to do so at the lowest possible bitrate [24, 64]. These two design approaches have led to two distinct design paradigms.

While constant bitrate coding practice has been prevalent to date, insufficient coding bitrates (due to transmission bandwidth and storage space constraints) and/or unregulated visual quality across spatiotemporal dimensions and views confront existing bitrate-driven design philosophy and approaches, which exacerbates visual traits of various spatial and temporal coding artifacts [71] and their spatial and/or temporal variations/fluctuations [11, 68], which markedly erode the quality of visual experience as intended by HD/UHD videos, and render ineffective and/or inefficient 3-D video presentations [64].

The current research and development in 3-DVC have been predominantly led and propelled by international standard activities associated with multi-view video coding (MVC) extensions to ITU-TH.264/ISO/IEC advanced video coding (H.264/AVC) [6, 21, 55] and ITU-TH.265/ISO/IEC high efficiency video coding (H.265/HEVC) [22, 34], respectively, with test conditions for 3-DV core experiments specified in [35] and subjective quality assessment methods standardized in [20]. Inter-view prediction is the central theme of MVC for reduction of spatiotemporal redundancies between the base view and its enhancement view(s), while MVC plus depth map (MVC+D) caters for depth image-based rendering and synthesis for a multitude of views beyond the conventional stereoscopic 3-D video [6, 34, 54, 55]. The R-D performance is a key criterion where the bitrate and the PSNR (peak signal-to-noise ratio) are used as the measures [37], and subjective assessments are based on the mean opinion score (MOS) using a five-level or an eleven-level absolute category rating (ACR) [14, 18, 20].

The supporting evidence was abundantly shown in the literature that these 3-DVC frameworks as the extensions to international video coding standards were effective in lifting the above-said objective or subjective picture quality significantly with a given (anchor) bitrate compared with their prior work [6, 34, 37]. What had been missing was the evidence to show whether these coding frameworks would be able to consistently deliver a prescribed or designated visual quality as discernible by human observers in terms of, e.g., JNND as well as JND levels [64] or visual distortion units (VDUs) [41].

Annoying visual experience due to various spatiotemporal video coding artifacts and their variations is, with all due respect, commonplace in applications of the existing single-view video coding standard implementations based on the rate-driven design approach [11, 71]. 3-DVC which follows the same design principle with unregulated or ineffectively controlled visual quality will inhere all the artifacts associated with its single-view counterpart. Furthermore, as mentioned previously, independent encoding of video frames of paired left and right views at unregulated quality levels or encoding the enhancement view (or inter-view prediction error image) with reference to encoded base view without adequate quality control will lead to perceivable difference between coded matching areas of the two views above the JND, visible disparity-compensated coding residual fluctuations in the enhancement view which differs from those in the base view, and variation or even loss of depth perception due to visual quality differences above a certain threshold between the two views when visual quality is not (or cannot be) adequately regulated [39, 64]. It begs the question whether perceptual 3-DVC based on visual quality-driven design [24, 64] and rate-perceptual-distortion optimization (RpDO) principle [23, 42, 64, 66] ought to be considered as a viable alternative to address these issues.

Perceptual coding techniques for single-view visual signal compression as classi-fied in [64] may be extended to 3-DVC either as a constant bitrate coder to maximize visual quality for a given bitrate or as a constant quality coder to minimize coding bitrate for a designated visual quality. While there had been numerous and ongoing reports on consistent performance gains in terms of bitrate savings and visual quality by perceptual coding of (single-view) visual signals, most of these perceptual coders did not maintain a constant visual quality. To date, successful visual quality-regulated still image coding results reported in the literature are limited to the JNND or perceptually lossless level [36, 48, 58]. There were very limited reports on human perception-based 3-DVC which extended human visual system (HVS) models used in single-view video coding to 3-DVC [39, 72] and/or experimented with inter-view masking effect in asymmetric 3-D stereoscopic video coding [39, 43], where a bitrate-driven coding design platform was often used as was the case with most of the perceptual (single-view) video coding with few exceptions [59, 66]. Considering that performance in the three key primary perceptual dimensions of 3-DVC, including picture quality, depth quality/acuity, and visual comfort [20], is significantly impacted by the visual quality of compressed base and enhancement views of 3-D video [39, 64], a visual quality-driven design approach may offer a viable, if not imperative, solution to quality-assured and sustainable 3-D visual communication services. The causes of a seemingly insurmountable impasse to the much-needed paradigm shift from bitrate-driven design to visual quality-driven design for 3-DVC, or visual signal coding in general, deserve a closer examination. While a number of issues have been highlighted in [60], they are summarized in the context of perceptual 3-DVC as follows.

3.3   Perceptual 3-D Video Coding

The past decade bore witness that digital 3-D full HD TV/video (3-DHDV) spearheaded technological advance in both viewing clarity and dimension, and showcased to the world the exhilarating visual experience brought about by the increase in video spatiotemporal resolutions (i.e., the HD) and provision of stereoscopic 3-D visualization [3]. It is noted that frame-compatible 3-DV formats by interleaving the two views after spatial down-sampling or temporal multiplexing left and right view frames after temporal down-sampling, which were used in initial 3-DV applications to reduce the bitrate required by simulcast at the expense of reduced spatial or temporal resolutions, are not considered as full-resolution 3-DHDV [54].

For 3-DV, MVC extensions to H.264/AVC [21] and H.265/HEVC [22], respectively, minimize both inter-frame (i.e., between frames of a video sequence) temporal redundancy by motion-compensated prediction and inter-view (i.e., between the two views of a 3-DV) redundancy by disparity-compensated prediction [54], demonstrating about 25 % and over 50 % bitrate savings for a given signal quality in terms of the PSNR compared with simulcast using H.264/AVC and H.265/HEVC, respectively [34, 37, 54]. However, HVS does not compute the mean squared error (MSE) or the PSNR [64]. Holding a constant MSE, or PSNR, or quantization step size or bitrate does not guarantee delivery of constant picture quality as perceived by the HVS, and minimizing MSE alone does not guarantee visual superiority [63, 64].

While overcoming a series of problems associated with the existing compression techniques for 3-DV as previously discussed calls for a long-waited paradigm shift from bitrate-driven design to visual quality-driven design approach to 3-DVC, the latter is not without its own challenges in at least three closely related research areas including 3-DVC, perceptual video coding (PVC), and quantitative estimation of visual picture quality levels as discernible by human observers. Successful resolution of challenging issues in these areas will not only improve quality of 3-DV services to secure their long-term viability, but also have significant theoretical and practical impacts on future deployments of, e.g., 3-DUHDV applications.

3.3.1 State of the Art in 3-DVC

There are three well-known approaches to 3-DVC [54], i.e., conventional MVC (including scalable video coding) [55], single-view video (SVV) plus depth map coding [33], and static or dynamic 3-D mesh compression [12]. MVC extensions of H.265/HEVC standard currently lead the state of the art for 3-DVC including full HD [22, 37], overtaking its predecessor MVC extensions of H.264/AVC [6, 21, 55], in terms of R-D optimization (R-DO) and visual quality improvement for a given bitrate under test conditions specified in [35] using standardized subjective quality assessment methods [18, 20]. The base view of 3-DV is coded independently using standard compliant coding to ensure backward compatibility to SVV and, therefore, its performance (or lack thereof) is referred to that of so-coded SVV [34, 37]. Subjective picture quality assessments are commonly based on the MOS using a five-level or an eleven-level ACR [14, 18, 20, 64].

The scalable video coding extension of the H.264/AVC standard was reported which utilized spatiotemporal scalability tools to scale low-resolution frame-compatible stereo 3-DV to full resolution. A wavelet-based scalable MVC was also reported in [10] with a comparable performance to that of an H.264/AVC standard MVC baseline implementation.

Three fundamental issues have not been fully, if at all, addressed by the above-reported 3-DVC research efforts. First and foremost, there had been no evidence to show that these 3-DVC frameworks would be able to consistently deliver a designated visual quality as discernible by human observers in terms of, e.g., JNND as well as JND levels [60, 64] or visual distortion units (VDUs) [41], but the opposite could be found or deduced [60, 64]. Decades have passed since digital video was introduced to the general public [24, 57], and the end users still cannot be provided with video services of a guaranteed visual quality [64]. More devastatingly when this bitrate-driven design approach is extended to 3-DV with inadequately regulated visual quality for the two 3-DV views, it may lead to viewing discomfort due to loss (and/or variations) of depth perception as previously illustrated by Fig. 3.2 [64]. Second, there is no evidence to show that spatiotemporal and signal quality scalabilities allowed within these existing frameworks can aid in RpDO to achieve uniform visual distortions across spatiotemporal dimensions and move from one JND level to the next with a required precision for any given or all spatiotemporal areas/regions of pictures with varying contents [60, 64]. Third, while the ACR has been widely used in subjective picture quality evaluations assuming the ground truth [14, 18, 20, 42], it is known that human perception and judgment in a psychophysical measurement task perform usually better in comparison tasks than casting an absolute rating [67]. The latter, including the ARC and the adjectival categorical judgment (ACJ) method [18], generates subjective test data with uncertainty and variations due to contextual effects [8] and varying experience and expectations of observers [60]. These three critical research issues associated with 3-DVC look to a visual quality-driven perceptual coding design approach for a solution.

3.3.2   Perceptual Video Coding

Perceptual picture coders have been reported extensively in the literature [63, 64] with a recent survey listing 16 representative perceptual video coders (PVCs) for SVV coding, 10 of which conform to existing international standards [26]. Successful approaches to PVC are classified as perceptual predictive coding, perceptual quantization, RpDO, and perception-based pre-, loop-, and post-filtering [64]. Compared with traditional waveform coding techniques, perceptual picture coding has clearly demonstrated superior visual quality performance at both perceptually lossless (e.g., a 5:1 bitrate reduction or more compared with information lossless JPEG 2000 coding for color images [36, 48]) and perceptually lossy levels where HVS-based quantization or perceptual distortion measures (PDMs) are used in

R-DO, achieving significant compression gains [5, 13, 26–28, 47, 49, 63, 64, 69, 70]. A maximum of 55.69 % in bitrate reduction was reported by an H.264/AVC-compliant PVC without degradation of visual quality compared with JM 14.2 (joint model) of the H.264/AVC [29]. A latest report on PVC compliant with the H.265/HEVC standard claimed a maximum of 49.10 % and an average 16.10 % bitrate reduction with negligible subjective quality loss [25]. Perceptually lossless compression approach has also been used for 2 and 4 K video coding [59]. However, there were very limited reports on perceptual 3-DVC, extending HVS models to 3-DVC [39, 72] or applying inter-view masking effect in asymmetric 3-D stereoscopic video coding [39, 43] with the most recent report on visually lossless coding of 3-D monochrome images using the JPEG 2000 framework [9].

A theoretical difficulty in perception-based hybrid video coding design is that available HVS models for RpDO are formulated using perceptual decomposition of an image or JND models of an image in spatial or transform domain, instead of a *residual* image which is the difference between the image and its estimate by intra- or inter-frame prediction [64]. A common practice is to discard prediction error/residual below the JND, while two techniques have been used to compress prediction error above the JND, i.e., off-set prediction errors [69] or normalize them by the JND [70]. Both techniques assumed that a linear or proportional scaling was extendable from threshold to suprathreshold region, which have not been proven, but counterexamples exist [41, 60, 62]. A theoretical framework of RpDO was formulated for 3-DVC with inter-frame and inter-view prediction using a PDM, solving an outstanding theoretical problem in perceptual coding of inter-frame and inter-view residual images for high-quality 3-DV applications [66].

Another theoretical challenge is that RpDO approach to visual quality-regulated 3-DVC design relies on design of PDM which captures 3-DVC distortions[1] corresponding to discernible levels by human visual perception, i.e., JND_i (or JND level i) for $i \in \mathbb{Z}^*$ where $\mathbb{Z}^*$ is the set of nonnegative integers and JND_0 is defined as JNND [41, 64, 66]. While existing PDMs successfully grade distortions on the ACR scales [15, 17, 63], it has been found that a number of well-known PDMs based on either threshold vision and/or suprathreshold vision models fail to predict discernible levels of distortions in terms of JND_i for $i \in \mathbb{Z}^*$ [60–62]. Although these PDMs are able to guide a PVC to spend its available bitrate budget to where it could achieve preferable visual quality on average, they are unable to accurately predict or maintain visual distortion at a given level across spatiotemporal domains in order to regulate visible differences between the two 3-DV views for effective inter-view masking of coding distortions and, at the same time, to avoid potential loss of depth perception. It is crucial for a PDM to predict accurately visual distortion to

---

[1]3-D video coding distortions refer to various artifacts and distortions in digitally coded 3-DV using mainstream coding and compression techniques [34, 37, 54, 64], including color bleeding across views, aliasing effects, compression induced cross talk, depth distortion, and cardboard artifacts due to insufficient level of quantization of depth information [30, 40], in addition to single-view coding artifacts [71].

determine *whether*, *to what extent*, and *if* inter-view masking effect of human stereo vision can be exploited in delivering quality-assured and robust 3-DV services [9, 18, 64].

Considerable temporal fluctuation distortion (also known as temporal flickering) exists which has plagued the wavelet transform-based video coding techniques [2, 60, 63]. While progress has been made in incorporating temporal vision models into video coding [25, 29, 50, 56, 59, 69, 70], 3-D HVS modelling for 3-D PVC [72], and HVS-based 3-D visual signal quality assessment measures/metrics [46], the existing models and implementations require further examination for visual quality-regulated 3-D PVC based on subjective test data in terms of JNDs [64].

3.4   Visual Quality-Regulated 3-D Video Coding

The difference between the classical R-D optimization (R-DO) and the RpDO was examined in [60, 62] where "0" measured by the MSE is replaced by the JNND (or JND_0 if you will) as the reference point while other nonzero MSE values by JNDs (or VDUs) on the distortion scale. In the RpDO-based constant quality coder design, maintaining a bitrate deems redundant or ineffective if, e.g., it is neither sufficient to guarantee a distortion level at the JNND for perceptually lossless coding nor necessary to achieve JND_1 (i.e., JND) for a perceptually lossy coding.

For a perceptual distortion measure (i.e., PDM) to regulate spatiotemporal and stereoscopic visual quality in RpDO-based 3-DVC, it is required to predict reasonably accurately discernible quality levels, i.e., JNDs, by human observers, to ensure a designated visual quality and to prevent severe visual discomfort due to, e.g., unstable or loss of depth perception [64]. A preliminary experimental investigation has been reported in [60, 62] to ascertain if various perceptual image distortion/quality metrics reported in recent years are able to consistently grade different images at various JND levels. Images were generated using an open-source JPEG 2000 coder at increasingly higher compression ratios for a total of 81 variations for each of 41 well-known test images, providing a range of test pictures which capture the transition points between JND levels. An image at JND_n is determined relative to the image at $JND_{(n-1)}$ for $n > 1$, whereas JND_1 is relative to the reference image, such that JND_2 is relative to JND_1 and JND_3 to JND_2, and so on [65]. Perceptual distortion or quality measures were computed for sets of images at JND_1, JND_2, JND_3, JND_4, and JND_5, respectively. Assuming small data samples with normal distribution, the lower and the upper bounds of the metrics' predictions were computed with the mean and the standard deviation of each metric using a 95 % confidence interval (CI), and prediction was considered inaccurate if the variation was such that most of the responses from a metric (i.e., >50 %) sat outside the 95 % CI range. It was found, as one might have expected,[2] that acceptance rates of these metrics were lower than 50 %, indicating that they were ineffective as a measure to predict discernible visual quality levels in terms of JNDs.

There is a need for alternative subjective test methods, where quality 3-D video test sequences are used to anchor the reference point with respect to visual quality scale tailored to applications to produce subjective data in terms of JNDs (or VDUs) which can be used in place of ACR-based subjective data to model and to parameterize PDMs for quality-regulated 3-DVC design and automated assessment methods.

As previously mentioned, another difficulty in visual quality-regulated 3-DVC design is related to the fact that the available perceptual distortion measures embedded in a video coder to date are often formulated using an image-based HVS model in either pixel or transform domain, instead of residual images which results from temporal or inter-view prediction [64]. JND thresholding [56] and JND adaptive inter-frame residual coding [69] have been reported where the prediction errors below the JND are eliminated from further coding and only residuals above the JND are adjusted with respect to the JND for further coding by a static or preset quantization strategy, which has been extended to perceptual 3-DVC in [72]. In either case, coding of inter-frame or inter-view prediction residual to a designated visual quality level has not been established and is subject to further examination.

3.5 A Theoretical Framework for Visual Quality-Regulated 3-D Video Coding

HVS model-based perceptual coding approach has demonstrated superior performance in terms of RpDO for still image coding and intra-frame coding of digital video against standard benchmarks, noticeably at perceptually lossless quality [36, 48, 58, 59, 64]. To address the issue with the lack of HVS models based on prediction residual image for video coding [64] and perceptual coding of 3-D video prediction error image [72], a perceptual 3-DVC framework with hybrid inter-frame and inter-view prediction is formulated using RpDO criterion [42, 66].

3.5.1 Perceptually Lossless 3-D Video Coding

Figure 3.3 illustrates a perceptually lossless 3-DVC framework with hybrid inter-frame motion- and inter-view disparity-compensated prediction, using the discrete

---

[2]Perceptual image distortion/quality metrics reported to date were mostly optimized or parameterized to fit subjective ACR data instead of JND levels, notwithstanding that some of these metrics were JND or threshold vision model based.

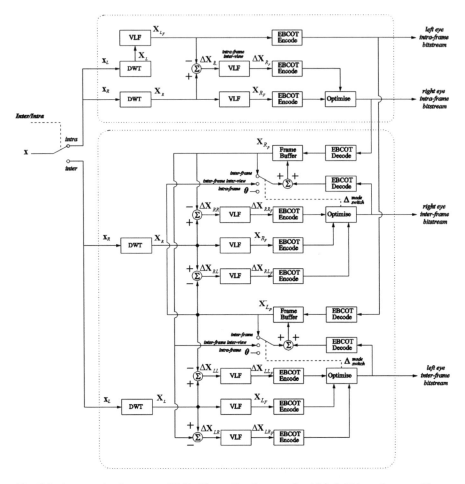

Fig. 3.3 An example of perceptual 3-D video coding framework with hybrid inter-frame and inter-view prediction [42, 66]. VLF – visually lossless filtering as defined in (3.1), EBCOT – embedded block coding with optimal truncation [52], Optimise – determining the minimum bitrate of the input coded streams as the output along with selected coding mode

wavelet transform (DWT) decomposition as an example, where the left view is the base view as reference and the right view is the enhancement view. In Fig. 3.3, function block "EBCOT Encode" is the embedded block coding with optimal truncation [52] and "Optimise" finds the minimum bitrate of the input coded streams to output along with selected coding mode. Function block "VLF" removes, via a visual filtering, as much data and psychovisual redundancies as possible from transform domain representations, minimizing the bitrate while keeping the PDM value equal to or below JND_0 (the JNND) threshold [42, 48, 58, 66], i.e.,

$$\mathbf{v}_F = T_{\mathrm{VLF}}\left(\mathbf{v}, \mathbf{v}_{\mathrm{ref}}\right)$$
$$\triangleq \arg\min_{\forall \mathbf{v}_f} R_{\mathrm{BIT}}\left\{\mathbf{v}_f \middle| \mathrm{PDM}_{\mathrm{JND}}\left(\mathbf{v}_O, \mathbf{v}_{v_f}\right) \le \mathrm{JND}_0\right\} \tag{3.1}$$

where the symbol "$\triangleq$" stands for "defined as," the operator T_{VLF} on a vector (matrix) variable $\mathbf{v}$ with a given reference vector variable $\mathbf{v}_{\mathrm{ref}}$ returns the visually filtered coefficient vector $\mathbf{v}_F$ which delivers the perceptually lossless compression using a perceptual distortion measure $\mathrm{PDM}_{\mathrm{JND}}$ for JND level prediction at the minimum bitrate, function $R_{\mathrm{BIT}}\left\{\cdot\right\}$ computes the bitrate, $\mathbf{v}_O$ is the original image with respect to $\mathbf{v}$ in the DWT domain, and $\mathbf{v}_{v_f}$ is a visually filtered coefficient vector as an approximation of $\mathbf{v}_O$ with respect to $\mathbf{v}$, and defined as

$$\mathbf{v}_{v_f} = T_f\left(\mathbf{v}, \mathbf{v}_{\mathrm{ref}}\right) \triangleq \mathbf{v}_f + \mathbf{v}_{\mathrm{ref}} \tag{3.2}$$

$\mathbf{v}_{v_f}$ is a visually filtered version of $\mathbf{v}$ by operator T_f, and $\mathbf{v}_{\mathrm{ref}}$ is zero in intra-frame coding with no reference (i.e., without intra-frame predictive coding). Given that $\mathbf{x}$ represents a given image and $\mathbf{X}$ is its DWT, here are examples of coding operations illustrated in Fig. 3.3.

3.5.1.1 Base View (Intra-Frame Mode)

Given the (base) left view image denoted by $\mathbf{x}_L$, its visually filtered DWT coefficient matrix $\mathbf{X}_{L_F}$ in intra-frame mode without predictive coding (i.e., $\mathbf{v}_{\mathrm{ref}} = 0$), which delivers the perceptually lossless compression at the minimum bitrate by enforcing the difference between DWT of $\mathbf{x}_L$ (i.e., $\mathbf{X}_L$) and its filtered DWT $\mathbf{X}_{L_f}$, using the perceptual distortion measure $\mathrm{PDM}_{\mathrm{JND}}$, at or below the threshold level defined by JND_0 at the minimum bitrate, is given by Eq. (3.1) as follows:

$$\mathbf{X}_{L_F} = T_{\mathrm{VLF}}\left(\mathbf{v}, \mathbf{v}_{\mathrm{ref}}\right)\Bigg|_{\substack{\mathbf{v} \triangleq \mathbf{X}_L \\ \mathbf{v}_{\mathrm{ref}} \triangleq 0}}$$
$$= \arg\min_{\forall \mathbf{X}_{L_f}} R_{\mathrm{BIT}}\left\{\mathbf{X}_{L_f} \middle| \mathrm{PDM}_{\mathrm{JND}}\left(\mathbf{X}_L, \mathbf{v}_{v_f}\right)\Big|_{\mathbf{v}_{v_f} \triangleq \mathbf{X}_{L_f}} \le \mathrm{JND}_0\right\}, \tag{3.3}$$

where JND_0 denotes the JNND, noting that $\mathbf{v} = \mathbf{v}_O = \mathbf{X}_L$ in Eq. (3.1), and in Eq. (3.2), and

$$\mathbf{v}_{v_f} = T_f\left(\mathbf{v}, \mathbf{v}_{ref}\right)\Bigg|_{\substack{\mathbf{v} \triangleq \mathbf{X}_L \\ \mathbf{v}_{ref} \triangleq 0}} = \mathbf{X}_{L_f} + 0. \tag{3.4}$$

EBCOT-encoded $\mathbf{X}_{L_F}$ is output as the perceptually lossless coded left view bit stream.

3.5.1.2   Enhancement View (Inter-View with Intra-Frame Mode)

For the inter-view coding of the (enhancement) right view where the left view $\mathbf{X}_{L_F}$ is used as a reference for inter-view disparity prediction, the minimum bitrate $R_{R\min}$ to encode the right view frame, $\mathbf{X}_R$, in the DWT domain to perceptually lossless level is given by

$$R_{R\min} = \min\left(R_{\mathrm{BIT}}\left(\mathbf{X}_{R_F}\right), R_{\mathrm{BIT}}\left(\Delta\mathbf{X}_{R_F}\right)\right) \tag{3.5}$$

where $\mathbf{X}_{R_F}$ is coded as shown in Eq. (3.3) by substituting $\mathbf{X}_R$ for $\mathbf{X}_L$, and

$$
\begin{aligned}
\Delta\mathbf{X}_{R_F} &= T_{\mathrm{VLF}}\left(\mathbf{v}, \mathbf{v}_{\mathrm{ref}}\right)\Big|_{\substack{\mathbf{v} \triangleq \Delta\mathbf{X}_R = \mathbf{X}_R - T_{dp}(\mathbf{X}_{L_F}) \\ \mathbf{v}_{\mathrm{ref}} \triangleq T_{dp}(\mathbf{X}_{L_F})}} \\
&= \arg\min_{\forall \Delta\mathbf{X}_{R_f}} R_{\mathrm{BIT}}\left\{\Delta\mathbf{X}_{R_f}\Big| \mathrm{PDM}_{\mathrm{JND}}\left(\mathbf{X}_R, \mathbf{v}_{v_f}\right)\Big|_{\mathbf{v}_{v_f} \triangleq T_{dp}(\mathbf{X}_{L_F}) + \Delta\mathbf{X}_{R_f}} \leq \mathrm{JND}_0\right\},
\end{aligned}
\tag{3.6}
$$

noting that $\mathbf{v}_O = \mathbf{X}_R$ in Eq. (3.1) and operator $T_{dp}(\cdot)$ represents disparity-compensated prediction [54] in the DWT domain using, e.g., a wavelet tree-based low-band-shift (LBS) method [38] with $\mathbf{X}_{L_F}$ as the reference.

3.5.1.3   Enhancement View (Hybrid Inter-Frame and Inter-View with Intra-Frame Mode)

For inter-coded right view frame with hybrid motion- and disparity-compensated prediction, the minimum bitrate $R_{PR\min}$ to encode the right view frame, $\mathbf{X}_R$, to perceptually lossless level is given by

$$R_{PR\min} = \min\left(R_{\mathrm{BIT}}\left(\mathbf{X}_{R_F}\right), R_{\mathrm{BIT}}\left(\Delta\mathbf{X}_{RR_F}\right), R_{\mathrm{BIT}}\left(\Delta\mathbf{X}_{RL_F}\right)\right) \tag{3.7}$$

where $\mathbf{X}_{R_F}$ is coded by substituting $\mathbf{X}_R$ for $\mathbf{X}_L$ in Eq. (3.3), and the inter-view predictive error with reference to the base view, $\mathbf{X}_{L_F}$, to be coded to perceptually lossless level is obtained as

$$
\begin{aligned}
\Delta\mathbf{X}_{RL_F} &= T_{\mathrm{VLF}}\left(\mathbf{v}, \mathbf{v}_{\mathrm{ref}}\right)\Big|_{\substack{\mathbf{v} \triangleq \Delta\mathbf{X}_{RL} = \mathbf{X}_R - T_{dp}(\mathbf{X}_{L_F}) \\ \mathbf{v}_{\mathrm{ref}} \triangleq T_{dp}(\mathbf{X}_{L_F})}} \\
&= \arg\min_{\forall \Delta\mathbf{X}_{RL_f}} R_{\mathrm{BIT}}\left\{\Delta\mathbf{X}_{RL_f}\Big| \mathrm{PDM}_{\mathrm{JND}}\left(\mathbf{X}_R, \mathbf{v}_{v_f}\right)\Big|_{\mathbf{v}_{v_f} \triangleq T_{dp}(\mathbf{X}_{L_F}) + \Delta\mathbf{X}_{RL_f}} \leq \mathrm{JND}_0\right\},
\end{aligned}
\tag{3.8}
$$

$\Delta\mathbf{X}_{RL_f}$ is a visually filtered prediction error, $\Delta\mathbf{X}_{RL}$, with respect to reference $T_{dp}(\mathbf{X}_{L_F})$ as defined in Eq. (3.6), and the inter-frame predictive error to be coded to perceptually lossless level with reference to a previously coded frame, e.g., $\mathbf{X}_{R_F}[i-1]$, is formulated as

$$
\begin{aligned}
\Delta\mathbf{X}_{RR_F} &= T_{\text{VLF}}(\mathbf{v}, \mathbf{v}_{\text{ref}}) \Big|_{\substack{\mathbf{v} \triangleq \Delta\mathbf{X}_{RR} = \mathbf{X}_R[i] - T_p(\mathbf{X}_{R_F}[i-1]) \\ \mathbf{v}_{\text{ref}} \triangleq T_p(\mathbf{X}_{R_F}[i-1])}} \\
&= \arg\min_{\forall \Delta\mathbf{X}_{RR_f}} R_{\text{BIT}} \left\{ \Delta\mathbf{X}_{RR_f} \Big| \text{PDM}_{\text{JND}}(\mathbf{X}_R, \mathbf{v}_{v_f}) \Big|_{\mathbf{v}_{v_f} \triangleq T_p(\mathbf{X}_{R_F}[i-1]) + \Delta\mathbf{X}_{RR_f}} \leq \text{JND}_0 \right\},
\end{aligned}
$$

$$(3.9)$$

$\Delta\mathbf{X}_{RR_f}$ is a visually filtered prediction error, $\Delta\mathbf{X}_{RR}$, with respect to reference $T_p(\mathbf{X}_{R_F}[i-1])$ which is a prediction of the current $\mathbf{X}_R[i]$ based on $\mathbf{X}_{R_F}[i-1]$. For instance, $T_p(\mathbf{X}_{R_F}[i-1]) = \mathbf{X}_{R_F}[i-1]$ represents a simple temporal differential pulse code modulation (DPCM) for the prediction whereas $T_p(\mathbf{X}_{R_F}[i-1])$ can also be implemented using a variety of motion-compensated prediction algorithms in the DWT domain [31, 32], including multi-resolution motion compensation (MRMC) [73], and wavelet tree-based MC using LBS method [38]. It is noted that in the example shown by Eq. (3.9), $\mathbf{v}_O = \mathbf{X}_R$ and $i \in I$, (I denotes a set of sequenced video frames), is the time index of the left and right video sequences which is omitted when all frames in the equation are at the same time instance.

Not all coding paths shown in Fig. 3.3 will be used depending on application constraints. For instance, backward compatibility requires $\Delta\mathbf{X}_{LR}$ and $\Delta\mathbf{X}_{LR_F}$ not to be used if the left view video is used in the base view decoding for single-view video applications. Using the (base) left view sequence coding path, this framework allows backward compatibility with perceptually lossless compression of single-view HD or UHD video for high-quality digital cinematic distribution applications [31, 32, 59]. A more generic intra-frame, inter-frame, and inter-view prediction-based coding such as biprediction and intra-prediction [21, 22, 54] can be similarly formulated and included in this framework [64].

The perceptually lossless coding scheme described above for 3-DVC is sufficient but may not be necessary according to the suppression theory of human 3-D visual perception (cf. Fig. 3.2) [43]. The performance of perceptually lossless 3-DV coder may be further improved by ensuring visually lossless quality performance for the base view while determining a reliable JND level for the enhancement view which exploits inter-view masking to deliver visually lossless quality of 3-DV to maximize compression ratio [64].

3.5.2 Visual Quality-Regulated 3-D Video Coding

Substituting JND_0 in Eq. (3.1) by a different JND level, i.e., JND_1, JND_2 and so on [60], visual quality-regulated 3-DVC or perceptual 3-DVC with perceptual distortion control can be achieved [64], where JND levels are mapped by predicted values using a PDM measure or profile [41, 48, 60]. User- or service-specific alternative quality/distortion scales may be defined in terms of JNDs or VDUs.

Visual decomposition transform is not limited to the DWT which is used herein with the EBCOT as a vehicle to demonstrate the theoretical framework built on previously published work [41, 48, 58, 59, 64].

The suppression theory of human visual perception of a 3-D scene from stereoscopic video states that if right and left views are transmitted and displayed with unequal spatial, temporal, and/or quality resolutions, the overall 3-D video quality is determined by the view with the better resolution [39, 43]. Therefore, visual quality adaptation of 3-DV may be achieved for a desirable constant perceived 3-DV quality by adaptation of the spatial and temporal resolution of the enhancement view with controlled perceptual distortion while encoding the base view at a designated visual quality controlled by RpDO with PDM_{JND}. Inter-view visual sensitivity and masking effects can be incorporated into formulation of PDM_{JND} and profiles [9] which are optimized or parameterized to fit JND-based subjective test data [65].

3.6   Summary

In this chapter, a theoretical framework is presented for visual quality-regulated 3-D video coding after a review of the state of the art and highlighting a number of critical issues and challenges in three areas closely related to the subject matter. For the visual quality-regulated 3-D video coding to deliver consistent visual quality services in 3-D or multi-view or free viewpoint TV communications, broadcasting, and entertainment applications, it requires further investigation and advancement in subjective quality assessment methods as well as 3-DV test materials to obtain subjective picture quality assessment data in steps of JNDs as reference or ground truth, design and optimization of perceptual distortion or quality measures to accurately estimate JND levels which are in reasonably good correlation with the subjective test data in terms of JNDs, and effective and efficient deployment of rate-perceptual-distortion optimization in perceptual 3-DVC where a JND-based perceptual distortion measure is used in place of the traditional mean squared error or perceptual distortion metrics which are parameterized or optimized for prediction of mean opinion scores on an absolute category rating scale.

References

1. Akar GB, Tekalp AM, Fehn C et al (2007) Transport methods in 3DTV—a survey. IEEE Trans Circ Syst Video Technol 17(11):1622–1630
2. Aravind R, Chen T (1996) On comparing MPEG-2 intraframe coding with a wavelet-based codec. In: Proceedings of international picture coding symposium, Melbourne, pp 353–357

3. Australian Communications and Media Authority (2010) Temporary trials of 3D TV and other emerging technologies—discussion paper. Commonwealth of Australia. Available via DIALOG. http://www.acma.gov.au/~/media/Broadcasting%20Spectrum%20Planning/Report/Report%20Digital%20TV/pdf/Temporary%20trials%20of%203D%20TV%20and%20other%20emerging%20technologies%20Discussion%20Paper%20September%202010. PDF. Accessed 1 Feb 2016

4. Berger T, Gibson JD (1998) Lossy source coding. IEEE Trans Inf Theor 44(10):2693–2723

5. Chandler DM, Hemami SS (2005) Dynamic contrast-based quantization for lossy wavelet image compression. IEEE Trans Image Process 14(4):397–410

6. Chen Y, Hannuksela MM, Suzuki T et al (2014) Overview of the MVC+D 3D video coding standard. J Vis Commun Image R 25(4):679–688

7. Cheng E, Burton P, Burton J et al (2012) RMIT3DV: pre-announcement of a creative commons uncompressed HD 3D video database. Proc QoMEX 2012,. pp 212–217

8. Corriveau P, Gojmerac C, Hughes B et al (1999) All subjective scales are not created equal: the effect of context on different scales. Signal Process 77(1):1–9

9. Feng H-C, Marcellin MW, Bilgin A (2015) A methodology for visually lossless JPEG2000 compression of monochrome stereo images. IEEE Trans Image Process 24(2):560–572

10. Garbas J-U, Pesquet-Popescu B, Kaup A (2011) Methods and tools for wavelet-based scalable multiview video coding. IEEE Trans Circ Syst Video Technol 21(2):113–126

11. Gong Y, Wan S, Yang K et al (2014) An efficient algorithm to eliminate temporal pumping artifact in video coding with hierarchical prediction structure. J Vis Commun Image R 25(7):1528–1542

12. Han S-R, Yamasaki T, Aizawa K (2007) Time-varying mesh compression using an extended block matching algorithm. IEEE Trans Circ Syst Video Technol 17(11):1506–1518

13. Hontsch I, Karam LJ (2002) Adaptive image coding with perceptual distortion control. IEEE Trans Image Process 11(3):214–222

14. International Telecommunication Union-Radiocommunication Sector (1998) Subjective assessment methods for image quality in high-definition television. Rec ITU-R BT.710-4, Geneva

15. International Telecommunication Union-Radiocommunication Sector (2004) Objective perceptual video quality measurement techniques for standard definition digital broadcast television in the presence of a full reference. Rec ITU-R BT.1683, Geneva

16. International Telecommunication Union-Radiocommunication Sector (2011) Studio encoding parameters of digital television for standard 4:3 and wide-screen 16:9 aspect ratios. Rec ITU-R BT.601-7, Geneva

17. International Telecommunication Union-Radiocommunication Sector (2011) Objective perceptual video quality measurement techniques for standard definition digital broadcast television in the presence of a reduced bandwidth reference. Rec ITU-R BT.1885, Geneva

18. International Telecommunication Union-Radiocommunication Sector (2012) Methodology for the subjective assessment of the quality of television pictures. Rec ITU-R BT.500-13, Geneva

19. International Telecommunication Union-Radiocommunication Sector (2012) Parameter values for ultra-high definition television systems for production and international programme exchange. Rec ITU-R BT.2020, Geneva

20. International Telecommunication Union-Radiocommunication Sector (2012) Subjective methods for the assessment of stereoscopic 3DTV systems. Rec ITU-R BT.2021, Geneva

21. International Telecommunication Union-Telecommunication Standardization Sector (2014) Advanced video coding for generic audiovisual services. Rec ITU-T H.264 (version 10), Geneva

22. International Telecommunication Union-Telecommunication Standardization Sector (2015) High efficiency video coding. Rec ITU-T H.265 (version 3), Geneva

23. Jayant NS, Johnston J, Safranek R (1993) Signal compression based on models of human perception. Proc IEEE 81(10):1385–1422

24. Jayant NS, Noll P (1984) Digital coding of waveforms: principles and applications to speech and video. Prentice-Hall, Upper Saddle River

25. Kim J, Bae S-H, Kim M (2015) An HEVC-compliant perceptual video coding scheme based on JND models for variable block-sized transform kernels. IEEE Trans Circ Syst Video Technol 25(11):1786–1800
26. Lee H-S, Ebrahimi T (2012) Perceptual video compression: a survey. IEEE J Sel Areas Commun 6(6):684–697
27. Leung R, Taubman D (2009) Perceptual optimization for scalable video compression based on visual masking principles. IEEE Trans Circ Syst Video Technol 19(3):309–322
28. Liu Z, Karam LJ, Watson AB (2006) JPEG2000 encoding with perceptual distortion control. IEEE Trans Image Process 15(7):1763–1778
29. Luo Z, Song L, Zheng S et al (2013) H.264/advanced video control perceptual optimization coding based on JND-directed coefficient suppression. IEEE Trans Circ Syst Video Technol 23(6):935–948
30. Meesters LMJ, IJsselsteijn WA, Seuntiens PJH (2004) A survey for perceptual evaluations and requirements of three-dimensional TV. IEEE Trans Circ Syst Video Technol 14(3):381–391
31. Mei L, Wu HR, Tan DM (2005) A new DWT/MC/DPCM video compression framework based on EBCOT. In: Proc SPIE 5960, visual communications and image processing 2005, Beijing, pp 59606I–1-9 http://dx.doi.org/10.1117/12.633509
32. Mei L, Wu HR, Tan DM (2008) A wavelet interlaced video coding framework. In: Proceedings of international conference consumer electronics, Las Vegas, Digest of Technical Papers 8.3-5. Doi: 10.1109/ICCE.2008.4588116
33. Müller K, Merkle P, Wiegand T (2011) 3-D video representation using depth maps. Proc IEEE 99(4):643–656
34. Müller K, Schwarz H, Marpe D et al (2013) 3D high-efficiency video coding for multi-view video and depth data. IEEE Trans Image Process 22(9):3366–3378
35. Müller K, Vetro A (2014) Common test conditions of 3DV core experiments. ITU-T SG16 WP3 and ISO/IEC JTC 1/SC 29/WG 11, Doc. JCT3VE1100, San Jose
36. Oh H, Bilgin A, Marcellin MW (2013) Visually lossless encoding for JPEG-2000. IEEE Trans Image Process 22(1):189–201
37. Ohm J-R, Sullivan GJ, Schwarz H et al (2012) Comparison of the coding efficiency of video coding standards—including high efficiency video coding (HEVC). IEEE Trans Circ Syst Video Technol 22(12):1669–1684
38. Park HW, Kim H-S (2000) Motion estimation using low-band-shift method for wavelet-based moving-picture coding. IEEE Trans Image Process 9(4):577–587
39. Perkins MG (1992) Data compression of stereo pairs. IEEE Trans Commun 40(4):684–696
40. Quan H-T, Le Callet P (2010) Video quality assessment: from 2D to 3D-challenges and future trends. Proceedings of ICIP 2010, Hong Kong, pp 4025–4028
41. Ramos MG, Hemami SS (2001) Suprathreshold wavelet coefficient quantization in complex stimuli: psychophysical evaluation and analysis. J Opt Soc Am A 18(10):2385–2397
42. Rao KR, Wu HR (2013) Perceptual coding of digital pictures, Tutorial. ICME 2013, San Jose
43. Saygih G, Gürler CG, Tekalp AM (2010) Quality assessment of asymmetric stereo video coding. Proceedings of ICIP 2010, Hong Kong, pp 4009–4012
44. Shannon CE (1948) A mathematical theory of communication. Bell Syst Tech J 27(3):379–423, 27(3):623–656
45. Shannon CE (1959) Coding theorems for a discrete source with a fidelity criterion. IRE Nat Conv Record 7:142–163
46. Su C-C, Moorthy AK, Bovik AC (2014) Visual quality assessment of stereoscopic image and video: challenges, advances, and future trends. In: Deng CW et al (eds) Visual signal quality assessment—issues of quality of experience. Springer International Publishing, New York, pp 185–212
47. Tan DM, Tan C-S, Wu HR (2010) Perceptual colour image coder with JPEG2000. IEEE Trans Image Process 19(2):374–383

48. Tan DM, Wu D (2013) Perceptually lossless and perceptually enhanced image compression system & method, Patent Int. Pub. No.: WO2013/063638 A2. WIPO, Geneva, Switzerland
49. Tan DM, Wu HR, Yu Z (2004) Perceptual coding of digital monochrome images. IEEE Signal Process Lett 11(2):239–242
50. Tang C-W (2007) Spatiotemporal visual considerations for video coding. IEEE Trans Multimedia 9(2):231–238
51. Tanimoto M (2012) FTV: free-viewpoint television. Signal Process Image Commun 27(6):555–570
52. Taubman DS (2000) High performance scalable image compression with EBCOT. IEEE Trans Image Proc 9(7):1158–1170
53. Vetro A (2010) Frame compatible formats for 3D video distribution. Proc ICIP 2010, Hong Kong, pp 2405–2408
54. Vetro A, Tourapis AM, Muller K et al (2011) 3D-TV content torage and transmission. IEEE Trans Broadcasting 57(2):384–394
55. Vetro A, Wiegand T, Sullivan GJ (2011) Overview of the stereo and multiview video coding extensions of the H.264/AVC standard. Proc IEEE 99(4):626–642
56. Wei Z, Ngan KN (2009) Spatio-temporal just noticeable distortion profile for grey scale image/video in DCT domain. IEEE Trans Circ Syst Video Technol 19(3):337–346
57. Wiegand T, Sullivan GJ (2011) The picturephone is here. Really. IEEE Spectrum 48(9):51–54
58. Wu D, Tan DM, Baird M et al (2006) Perceptually lossless medical image coding. IEEE Trans Med Imag 25(3):335–344
59. Wu D, Tan DM, Wu HR (2010) Perceptual coding at the threshold level for the digital cinema system specification. In: Frossard P, et al., (eds), Proc ICME 2010, Singapore, pp 796–801
60. Wu HR (2014) Introduction—state of play and challenges of visual quality assessment. In: Deng CW et al (eds) Visual signal quality assessment—issues of quality of experience. Springer International Publishing, New York, pp 1–30
61. Wu HR (2015) QoE subjective and objective evaluation methodologies. In: Chen CW et al (eds) Multimedia quality of experience (QoE): current status and future requirements. Wiley, New York, pp 123–148
62. Wu HR, Lin W, Ngan KN (2014) Rate-perceptual-distortion optimization (RpDO) based picture coding—issues and challenges. In: Proceedings of 19th international conference digital signal process, Hong Kong, pp 777–782
63. Wu HR, Rao KR (eds) (2006) Digital video image quality and perceptual coding. CRC Press, Boca Raton
64. Wu HR, Reibman A, Lin W et al (2013) Perceptual visual signal compression and transmission. (invited paper). Proc IEEE 101(9):2025–2043
65. Wu HR, Tan DM (2015) Subjective and objective picture assessment at supra-threshold levels. In: Proceedings of picture coding symposium, Cairns, May–June 2015. pp 312–316
66. Wu HR, Tan DM, Wu D (2011) A theoretical framework for perceptually lossless coding of stereo 3-D video. Proceedings of 13th IASTED international conference signal and image proc (SIP 2011), Dallas, pp 50–55
67. Wu HR, Yu Z, Qiu B (2002) Multiple reference impairment scale subjective assessment method for digital video. Proceedings of 14th international conference digital signal process, Santorini vol 1, pp 185–189
68. Yang K, Wan S, Gong Y et al (2016) A novel SAO-based filtering technique for reduction in temporal flickering artifacts in H.265/HEVC. Circuits Syst Signal Process. http://link.springer.com/article/10.1007/s00034-016-0251-5
69. Yang XK, Lin W, Lu Z et al (2005) Motion-compensated residue preprocessing in video coding based on just-noticeable-distortion profile. IEEE Trans Circ Syst Video Technol 15(6):742–752
70. Yang XK, Ling WS, Lu ZK et al (2005) Just noticeable distortion model and its applications in video coding. Signal Process Image Commun 20(7):662–680

71. Yuen M, Wu HR (1998) A survey of hybrid MC/DPCM/DCT video coding distortions. Signal Process 70(3):247–278
72. Zhang L, Peng Q, Wang Q-H et al (2011) Stereoscopic perceptual video coding based on just-noticeable-distortion profile. IEEE Trans Broadcast 57(2):572–581
73. Zhang Y-Q, Zafar S (1992) Motion-compensated wavelet transform coding for color video compression. IEEE Trans Circ Syst Video Technol 2(3):285–296

Chapter 4
Recent Advances on 3D Video Coding Technology: HEVC Standardization Framework

Dragorad A. Milovanovic, Dragan Kukolj, and Zoran S. Bojkovic

Abstract 3D video is emerging media extension of conventional 2D video into third dimension adding depth sensation and resolving 2D viewing ambiguity. Primary usage scenario for 3D video is to support 3D video applications, where 3D depth perception of a visual scene is provided by a 3D display system. Multiview-plus-depth (MVD) is visual representation and coding format which takes 3D geometry information of acquisition system in the form of distance information. Applications require transmission of jointly encoded multiple synchronized video signals that show the same 3D scene from different viewpoints. Advances in multi-camera arrays and display technology enable new applications for 3D video. It is clear that these applications need to be based on well-defined and -documented technical standards. Recent advances and challenges in development of the 3D video formats and associated coding technologies are summarized in this chapter with focus on undergoing MPEG/ITU standardization framework for 3D extensions of HEVC high-efficiency video encoder. Research on coding efficiency improvement and complexity reduction of 3D-HEVC reference encoder implementation are outlined.

4.1 Introduction

Over the past 25 years, significant progress has been made in the coding and transmission of digital video. Three-dimensional digital video (3DV) signal processing technology has significantly affected the multimedia on Internet. The MPEG (*Moving Picture Coding Experts Group*) was established in January 1988 with the mandate to develop international standards for compression, decompression, processing, and coded representation of moving pictures, audio, and their combination, in order to satisfy a wide variety of applications. The ISO standards produced

D.A. Milovanovic • Z.S. Bojkovic
University of Belgrade, Studentski Trg 1, Belgrade 11000, Serbia
e-mail: dragoam@gmail.com

D. Kukolj (✉)
University of Novi Sad, Trg Dositeja Obradovica 6, Novi Sad 21000, Serbia
e-mail: dragan.kukolj@rt-rk.uns.ac.rs

© Springer Science+Business Media New York 2017
A. Kondoz, T. Dagiuklas (eds.), *Connected Media in the Future Internet Era*,
DOI 10.1007/978-1-4939-4026-4_4

Fig. 4.1 Standardization scope of digital video codec is indicated by *dashed boxes*: only the syntax and semantics of the bit stream and its decoding are defined

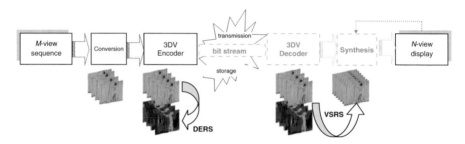

Fig. 4.2 Standardization scope of 3D video codec is indicated by *dashed boxes*: MPEG software DERS (Depth Estimation Reference Software) generates multiview+depth sequence, MPEG software VSRS (View Synthesis Reference Software) reconstructs multiview+depth and synthesizes *N*-views to display ($N \geq M$)

by MPEG are published in the last stage of a long process that starts with the proposal of new work within a committee, and continue through *competitive* and *cooperative* phases (Fig. 4.1). The evaluation of coding techniques is performed based on their performance (both objectively and by formal subjective testing), efficiency with respect to software/hardware implementation, and feasibility of system architectures.

ISO/IEC JTC1/SC29/WG11 *Moving Picture Experts Group* (**MPEG**) and ITU-T Study Group 16 *Video Coding Experts Group* (**VCEG**) are committees responsible for the development of video coding standards. These committees have jointly developed the widely deployed advanced video coding (**AVC**) and high-efficiency video coding (**HEVC**) standards. They are working on 3D extensions of these standards under the *Joint Collaborative Team on 3D Video* (**JCT-3V**), which was established in July 2012. The 3D video extensions (3D-HEVC, MV-HEVC, 3D-AVC, MVC+D) support the improved coding of stereoscopic and multiview video and facilitate advanced 3D capabilities such as view rendering through the use of depth maps (Fig. 4.2). Support for multiview enables representation of video content with multiple camera views and optional auxiliary information. Support for 3D enables *joint* representation of *video content* and *depth information* with multiple camera views.

The 3DV format targets two specific application scenarios:

- Enabling stereo devices to cope with varying display types and sizes, and different viewing preferences. This includes the ability to vary the baseline distance for stereo video to adjust the depth perception, which could help to avoid fatigue and other viewing discomforts.
- Support for high-quality auto-stereoscopic displays, in such a way that the new format enables the generation of many high-quality views from a limited amount of input data, e.g., stereoscopic video and respective depth maps.

Requirements for 3DV data format are as follows:

- **Video data.** The uncompressed data format shall support stereo video, including samples from left and right views as input and output. The source video data should be rectified to avoid misalignment of camera geometry and colors. Other input and output configurations beyond stereo should also be supported.
- **Supplementary data**. Supplementary data shall be supported in the data format to facilitate high-quality intermediate view generation. Examples of supplementary data include depth maps, reliability/confidence of depth maps, segmentation information, transparency or specular reflection, occlusion data, etc. Supplementary data can be obtained by any means.
  - **Metadata.** Metadata shall be supported in the data format. Examples of metadata include extrinsic and intrinsic camera parameters, scene data, such as near and far plane, and others.

Requirements for compression of 3DV data format are as follows:

- **Compression efficiency.** Video and supplementary data should not exceed twice the bit rate of state-of-the-art compressed single video. It should also be more efficient than state-of-the-art coding of multiple views with comparable level of rendering capability and quality.
- **Synthesis accuracy.** The impact of compressing the data format should introduce minimal visual distortion on the visual quality of synthesized views. The compression shall support mechanisms to control overall bitrate with proportional changes in synthesis accuracy.
- **Backward compatibility.** The compressed data format shall include a mode which is backward compatible with existing MPEG coding standards that support stereo and mono video. In particular, it should be backward compatible with MVC.
- **Stereo/mono compatibility**. The compressed data format shall enable the simple extraction of bit streams for stereo and mono output, and support high-fidelity reconstruction of samples from the left and right views of the stereo video.

Requirements for rendering of 3DV data format areas are as follows:

- **Rendering capability.** The data format should support improved rendering capability and quality. The rendering range should be adjustable.

- **Low complexity.** The data format shall allow real-time decoding and synthesis of views, required by any N-view display, with computational and memory power available to devices at the consumer electronics level.
- **Display types.** The data format shall be display independent. Various types and sizes of displays, e.g., stereo and auto-stereoscopic N-view displays of different sizes with different number of views, shall be supported. The data format shall be adaptable to the associated display interfaces.
- **Variable baseline.** The data format shall support rendering of stereo views with a variable baseline.
- **Depth range.** The data format should support an appropriate depth range.
- **Adjustable depth location.** The data format should support display-specific shift of depth location, i.e., whether the perceived 3D scene (or parts of it) is behind or in front of the screen.

Therefore, new coding methods are required for 3DV coding, which decouple the production and coding format from the display format. The primary goal of coding method is to *optimize* coding efficiency. *Coding efficiency* is the ability to minimize the bit rate necessary for representation of video content to reach a given level of video quality or, as alternatively formulated, to maximize the video quality achievable within a given available bit rate (Fig. 4.3).

The 3DV extensions based on the HEVC are developed jointly by MPEG and ITU-T for multiview video data with associated depth maps (MVD) coding for the highest compression efficiency. The 3D-HEVC base view is fully compatible with HEVC in order to extract monoscopic video, while the coding of dependent views and depth maps utilizes additional tools. HEVC *video coding layer* design is based on conventional block-based motion-compensated hybrid video coding concepts (Fig. 4.4). In HEVC, the main goal was to achieve a compression gain higher when compared to the second-generation video coding standard AVC at the same video quality. HEVC is targeted at next-generation ultra-HD (4/8K pixels per line) displays.

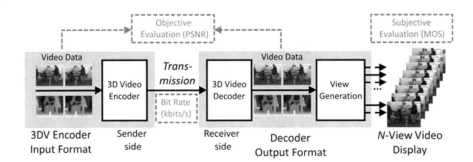

Fig. 4.3 Overall evaluation of different 3DV coding methods with compression and view generation methods (quality and data rate measurements are indicated by *dashed boxes*)

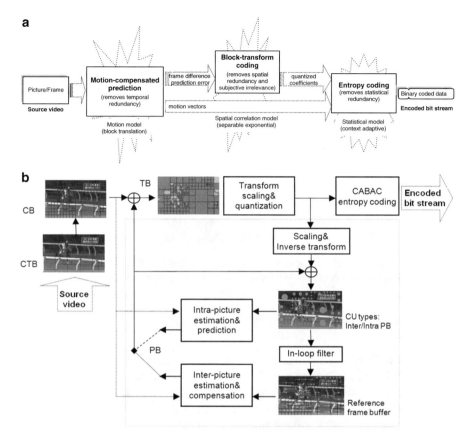

Fig. 4.4 Principles of digital video coding: (**a**) models exploit statistical redundancy of image and subjective irrelevance of viewer, (**b**) HEVC block-based hybrid MC prediction + TC transform coding

4.2 Three-Dimensional Video Formats and Associated Compression Technology

Efficient representation of three-dimensional (3D) video data is very closely involved with the other components of a system: content production, transmission, rendering, and display. It also has a significant impact on the overall performance of the system, including bandwidth requirement and end-user visual quality, as well as constraints such as backward compatibility with display equipment and transmission infrastructure. In this context, standardization is the key to guarantee interoperability and support mass deployment.

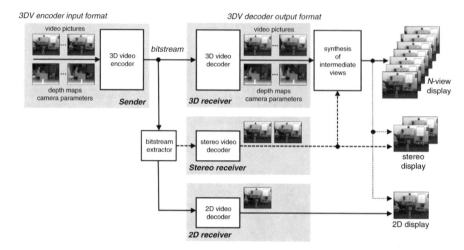

Fig. 4.5 Overview of the system structure and the data formats for encoding, transmission, decoding, rendering, and display of multiview video and associated depth maps

A variety of 3D video representations are available in the current ecosystem (Fig. 4.5):

- **Stereoscopic 3D** (S3D) video is the simplest and most widely used representation. It is based on the principle of *stereopsis*, in which two 2D views (L, R) with a disparity (D) are, respectively, received by the left and right eyes of an observer. The resulting binocular disparity is then exploited by the human visual system (HVS) to create a perception of depth in 3D scene.
- **Multiview video** (MVV) is a straightforward extension of the S3D representation; several texture videos are acquired in a *synchronous* manner by a system of cameras.
- **Multiview video-plus-depth** (MVD) is augmented with an extra channel conveying depth information. Depth maps (M) result in a display-independent representation that enables synthesis of a N ($N>M$) of views. The two sequences, *video texture* and *depth maps*, can then be encoded and transmitted independently. Alternatively, texture and depth can be jointly encoded, to exploit the redundancies between them, resulting in better coding performance.

The most important 3D video standardized codecs and associated formats are as follows:

- **Simulcast** is the simple independent coding (AVC/HEVC) and transmission of views. In addition, no synchronization between views is required. However, simulcast is not optimal in terms of rate-distortion efficiency because the correlation between cameras is not exploited.
- Multiview video coding (**MVC**) is AVC extension that exploits redundancy between views using inter-camera prediction to reduce required bit rate.

- Multiview video + depth coding (**3DV**) is in the current focus of MPEG standardization (MVC extension MVC+D, AVC-compatible extension 3D-AVC; HEVC extensions MV-HEVC and 3D-HEVC). Two major objectives are targeted: to support advanced stereoscopic display processing and to improve support for high-quality auto-stereoscopic multiview displays. It disconnects the video representation/coding from the captured video representation, and the displayed video representation.

For the sake of completeness, the other standardized 3D video formats are listed as follows:

- MPEG-2 Multiview Profile (MVP) uses scalable coding tools in transmission of two stereoscopic video signals inside an MPEG-2 transport stream, and guarantees backward compatibility with the MPEG-2 main profile.
- MPEG-C Part 3 specifies high-level syntax that allows an MPEG-2/AVC decoder to interpret two video streams correctly as texture and depth data inside an MPEG-2 transport stream.
- MPEG-4 Part 2 Multiple Auxiliary Components (MAC) specify a tool for coding video-plus-depth data.
- MPEG-4 Part 10 MVC multi-resolution frame-compatible stereoscopic video coding (MFC) specifies stereo interleaving (spatial/temporal multiplexing) formats, SEI (supplemental enhancement metadata) signaling massages for frame packing, as well as MFC+D enhancement for stereoscopic video coding with depth information.

4.2.1 3D-HEVC System Structure

3DV extensions based on the HEVC are developed jointly by MPEG and ITU-T for multiview video data with associated depth maps (MVD) coding for the highest compression efficiency. The 3D-HEVC base view is fully compatible with HEVC in order to extract monoscopic video, while the coding of dependent views and depth maps utilizes additional tools (Fig. 4.6). A subset of this 3DV coding extension includes MV-HEVC simple multiview extension, utilizing the same design principles of MVC in the AVC framework (providing backward compatibility for monoscopic decoding). MV-HEVC and 3D-HEVC extension are available as a final standards by mid-2014 and 2015, respectively. Additionally, it is planned to develop a suite of tools for scalable coding, where both view scalability and spatial scalability would allow for backward-compatible extensions for more views.

The system structure of 3D-HEVC is described as follows. The video pictures and depth maps are coded by **access units** as illustrated in Fig. 4.7. An access unit includes all video pictures and depth maps at the *same* time instant. The video picture and depth map corresponding to a particular camera position are indicated by a view identifier (`viewId`). The view identifier is also used for specifying the coding order. The view with view identifier equal to 0 is also referred to as the **base**

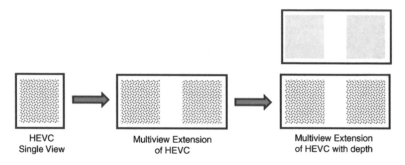

Fig. 4.6 3DVC extensions of HEVC coding standard: MV-HEVC supports MVV format, and 3D-HEVC supports MVD format

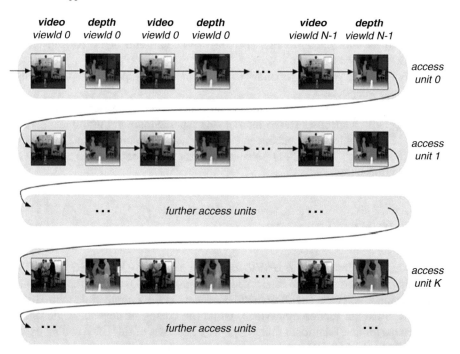

Fig. 4.7 3D-HEVC access unit structure

view or the **independent view** and is coded independently of the other views using a conventional HEVC video coder. The other views are referred to as **dependent views** and they can be coded with additional coding tools in 3D-HEVC.

4.2.2   3D-HEVC Encoding Process

The coding structure in 3D-HEVC includes three basic units, identical to that in HEVC: coding unit (**CU**), prediction unit (**PU**), and transform unit (**TU**). A picture is divided into a set of coding tree units (**CTUs**). The CTU is equivalent to a macroblock in H.264/AVC.

- The CU is represented as the leaf node of a *quadtree partitioning* of the CTU. It is a basic unit with a square shape which is associated with a **prediction mode**: intra, inter, or SKIP. A CTU may contain only one CU or may be split into four smaller CUs, and each CU could be recursively split into smaller CUs until the predefined splitting limitation is reached.
- A PU is a basic unit for prediction and has its root at the CU level. The shape of a PU is not necessarily square. Each CU may contain one, two, or four PUs according to the partition mode. The eight partition modes that can be used for an inter-coded CU are shown in Fig. 4.8. Only the PART_2Nx2N and PART_NxN

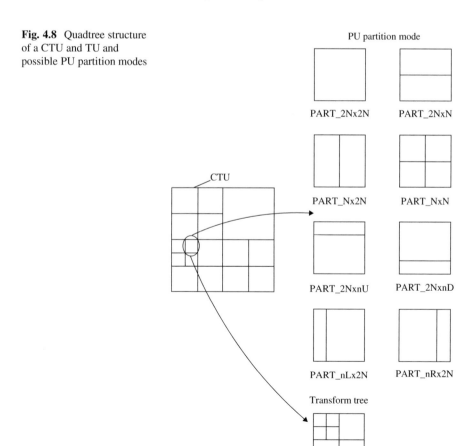

Fig. 4.8 Quadtree structure of a CTU and TU and possible PU partition modes

partition modes are used for an intra-coded CU. For both inter-coded CU and intra-coded CU, the partition mode PART_NxN can be allowed only when the corresponding CU size is equal to the minimum CU size.

- A TU is another basic unit with a square shape for transform and quantization. Multiple TUs within a CU form a quadtree structure called Residual QuadTree (RQT).

The 3D-HEVC encoder tests all the **coding modes** (up to 20 different modes, i.e., inter/merge/skip2N_2N, inter/merge 2N_N, inter/mergeN_2N, inter/merge N_N, inter/merge 2N_nU, inter/merge2N_nD, inter/mergenL_2N, inter/mergenR_2N, intra2N_2N, intraN_N, and intra PCM) for each CU and selects the mode with the least **RD cost**. Furthermore, each CU could be recursively split into four sub-CUs and the coding mode of each sub-CU is again determined by examining the RD cost of all the coding modes. Whether the CU should be further split or not is also decided by comparing the RD cost of the CU to the summation of the RD costs of the four sub-CUs. The motion estimation (**ME**) and the computation of the RD cost for each CU are the most computationally intensive parts.

The independent view, which is also referred to as the **base view**, is coded by a conventional HEVC codec. For dependent views, additional tools exploiting *inter-component* correlations have been integrated into 3D-HEVC (Fig. 4.9):

- To share the previously encoded texture information of reference views, the **disparity-compensated** prediction (**DCP**) has been added as an alternative to motion-compensated prediction (MCP).
- The **inter-view motion** prediction is employed to predict the motion information for the current block from the previously encoded motion information in the reference views.
- The **residual signal** of the current block can also be predicted from the residual signal of the corresponding block in the reference views.
- Backward **view synthesis prediction** (VSP) is a technique that exploits inter-view redundancies, in which a synthesized signal is used as a reference to predict the current picture.
- For the depth component, among all the above additional tools, only DCP is enabled. However, some new intra-prediction **depth modeling modes** (DMC) and the **motion parameter inheritance** (MPI) mode are added.

4.3   HEVC Standardization Framework

High-efficiency video coding (HEVC) is the current joint video coding standardization project of the ITU-T Video Coding Experts Group (ITU-T Q.6/SG 16) and ISO/IEC Moving Picture Experts Group (ISO/IEC JTC 1/SC 29/WG 11). The *Joint Collaborative Team on Video Coding* (JCT-VC) was established to work on this project. The scope of this group was extended to continue working on format range extensions (RExt), scalable HEVC (SHVC), and screen content coding (SCC) as

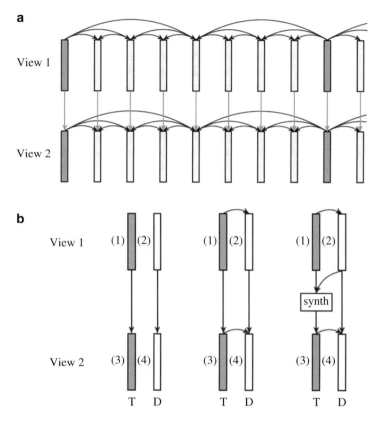

Fig. 4.9 Examples of 3D video prediction structures: (**a**) MVC inter-view prediction (view 1 is base/independent view), (**b**) MV-HEVC-independent coding of video texture (T) and depth maps (D), inter-component prediction within the same view, and BVSP view-synthesis prediction

extensions of HEVC. The *Joint Collaborative Team on 3D Video Coding Extension Development* (JCT-3V) was established to work on multiview and 3D video coding extensions of HEVC.

The main steps of HEVC technical developments are organized in four phases:

1. The HEVC first base specification finalized in 2013.
2. *Fidelity range extensions* (FRExt), *scalable video coding* (SHVC), and *multiview video coding* (MV-HEVC) extensions finalized in 2014.
3. 3D video coding (3D-HEVC) extension finalized in 2015.
4. *SCC* extensions will be included in the fourth version of HEVC, which is expected to be finalized in 2016.

Where as the first three developments mainly targeted compression performance for consumer and professional uses, SHVC and MV/3D video coding have provided additional functionality such as variable rate access at the bit stream level and support for multiple camera inputs in combination with efficient compression.

After finalization of the HEVC base specification, JCT-VC continued to work on extensions.

1. The *format range extension* (RExt) provides tools to support 4:0:0, 4:2:2, and 4:4:4 color spaces and additional bit depths. RExt is included in the second version of HEVC, which has been finalized in October 2014.
2. Already during the initial phase of HEVC, *multi-layer extensions* were planned and the proper hooks were included into the base specification. The *scalability extension* of HEVC (SHVC) provides support for spatial, SNR, and color gamut scalability. It has been designed as a high-level syntax-only extension to allow reuse of existing decoder components. SHVC is included in the second version of HEVC, which has been finalized in October 2014.

    The JCT-3V was established to work on multiview and 3D video coding extensions of HEVC and other video coding standards. The *multiview extension* of HEVC (MV-HEVC) provides support for coding multiple views with inter-layer prediction. It was designed as a high-level syntax-only extension to allow reuse of existing decoder components. MV-HEVC is included in the second version of HEVC, which was finalized in October 2014.
3. The *3D extension* of HEVC (3D-HEVC) provides increased coding efficiency by *joint coding* of texture and depth for advanced 3D displays. 3D-HEVC is included in the third version of HEVC, which was finalized in February 2015.
4. The SCC extensions will improve compression capability for video containing a significant portion of rendered (moving or static) graphics, text, or animation rather than (or in addition to) camera-captured video scenes. Example applications include wireless displays, remote computer desktop access, and real-time screen sharing for videoconferencing. SCC will be included in the fourth version of HEVC, which is expected to be finalized in February 2016.

JCT-VC adopted an open standardization approach in the development of specifications. All inputs and contributions to a JCT-3V meeting are made by documents which are registered in a publicly accessible document repository. A set of deliverables, which turn to become normative or remain to be supplemental in their final form, are also publicly accessible. These comprise the **specification text** itself, the **reference software**, a **conformance specification**, and the **test model**. Furthermore, a **verification report** is produced which documents and demonstrates the achieved performance.

- **Draft specification** is developed as a *working draft document* or *draft amendment*, depending on the state of the working phase. Since this document represents the current state of the main deliverable of the group, it has highest priority. A new version of the draft text is released after every meeting, integrating the adoptions of the meeting. While the specification of the first version of 3D-HEVC has been finalized, ongoing JCT-3V work on maintenance and extensions is reflected in corresponding specification drafts. Depending on the scale of the introduced changes, they may be published as an *amendment* or as a *new edition* of the standard. While amendments only include the applicable changes and extensions of the specification, a new edition would imply the publication of a complete integrated version of the specification text.

- **Test model document** is maintained aligned with draft specification. In distinction from the original HM reference software for HEVC, the 3D-HEVC reference software is referred to as **HTM**. The text describes the encoder control and algorithms implemented in the reference software which implements the reference decoder and a rate-distortion optimized encoder. The document aids analysis of the reference software, including the integrated normative tools. By describing the encoder decisions for application of the specified tools, the test model text serves as a tutorial example on how to implement an encoder control for the tool set in the specification.
- **Reference software** implements the decoding process as specified in the (draft) specification and an example encoder which generates bit streams complying to the specification. A new version of the **HTM software** is released after every meeting, integrating the adoptions of the meeting. In the development phase, the reference software specifically serves as the platform to test and analyze proposed tool changes. *Simulations* which are performed using the reference software confirm the expected rate-distortion performance along the development of the specification draft. The software **reference decoder** can be used by encoder manufacturers for testing if their encoded bit streams comply to the specification. Since the reference software does not necessarily include all restrictions specified in the text, successful decoding of a bit stream by this software may give a good indication but not a final proof for compliance. The reference software is maintained and developed to meet the goal of compliance as closely as possible. The **reference encoder** provides a rate-distortion optimized implementation, which aids in comparing the performance impact of tools in the context of the reference model. It should be noted that the reference software commonly does not include sophisticated rate control for real encoding tasks nor does it include significant error concealment in the decoder in the case that, e.g., corrupted bit streams are fed to the decoder. Such tools are up to the implementers of encoders and decoders for their respective target applications.
- **Conformance specification** is developed to provide means to manufacturers of encoders and decoders to test their product for compliance to the specification text. An important means for conformance testing of decoders is a suite of conformance bit streams which are generated by JCT-3V. These bit streams are designed to include a test set as complete as possible for all tools included in the specification. With the approval of the final version of the specification text, the design task for the conformance specification is to approach completeness as much as possible.
- **Core experiment** is the regular process for a tool to be adopted into the draft specification. While the proponent reports the test performance results of the addition or modification of coding tool, the most important task is on the core experiment participants who provide a *cross-check* of the proposed technology. Conceptually, a successful core experiment can be considered as the last step before adopting a proposal into the draft specification. However, the successful

Table 4.1 Publication of ITU-T Rec. H.265 and ISO/IEC international standard MPEG-H

ITU-T Rec. H.265.1 *High efficiency video coding*	13.04.2013
Annex A *Profiles, tiers and levels* (3 profiles)	
Annex B *Byte stream format*	
Annex C *Hypothetical reference decoder*	
Annex D *Supplemental enhancement information*	
Annex E *Video usability information*	
ITU-T Rec. H.265.2 *Reference software for ITU-T H.265 high efficiency video coding*	10/2014
ITU-T Rec. H.265.3 *Conformance specification for ITU-T H.265 high efficiency video coding*	10/2014
ISO/IEC 23008-2 :2013 *High efficiency video coding*	01.12.2013
ISO/IEC 23008-5:2013 *Reference software for HEVC*	16.10.2013
ISO/IEC 23008-8:2013 *Conformance specification for HEVC*	16.10.2013
ITU-T Rec. H.265 *High efficiency video coding*	29.10.2014
Annex F *Common syntax, semantics and decoding process for multi-layer video coding extensions*	
Annex G *Multiview coding* (multiview main profile)	
Annex H *Scalable high efficiency video coding*	
H.265.2 *Reference software for HEVC v2 (RExt, SHVC, MV-HEVC)*	10/2015
H.265.3 *Conformance specification for HEVC v2 (RExt, SHVC, MV-HEVC)*	10/2015
ISO/IEC 23008-2 :2015 *High efficiency video coding*	01.05.2015
ISO/IEC 23008-5:2015 *Reference software for HEVC*	15.04.2015
ISO/IEC 23008-8:2015 *Conformance specification for HEVC*	15.04.2015
ITU-T Rec. H.265 *High efficiency video coding*	29.04.2015
Annex I *3D High efficiency video coding* (3D main profile)	
ITU-T Rec. H.265.2 *Reference software for HEVC v3 (3D-HEVC)*	10/2015
ITU-T Rec. H.265.3 *Conformance specification for HEVC v3 (3D-HEVC)*	10/2015

conduction of a core experiment does not imply guaranteed adoption. Studies of changes in structures above the coding layer (the high-level syntax) do not easily allow for verification of the benefit of proposed changes. In such cases, assessment by qualified experts is obligatory.

VCEG and MPEG jointly developed the three versions of high-efficiency video coding specifications and published Recommendation ITU-T H.265 and ISO/IEC 23008-2 International Standard in a technically aligned manner (Table 4.1):

- The first edition refers to the first approved 04/2013 version of this Recommendation | International Standard. *Annex A* specifies profiles, tiers, and levels as restrictions on the bit streams and hence limits on the capabilities needed to decode the bit streams.
- The second edition approved 10/2014 refers to the integrated text containing format range extensions, scalability extensions, multiview extensions, and additional supplement enhancement information. *Annex G* specifies syntax, semantics, and decoding processes for multiview high-efficiency video coding (MV-HEVC). This annex also specifies profiles (Multiview Main), tiers, and levels for multiview high-efficiency video coding.

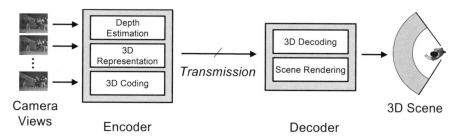

Fig. 4.10 Multiview + depth video processing chain

- The third edition approved 04/2015 refers to the integrated text containing 3D extensions. *Annex I* contains support for *3D high-efficiency video coding* (3D-HEVC), specifying a syntax and associated decoding process for efficient coding of video textures and depth maps for 3D video applications. One additional profile is defined in this revision, the 3D Main profile.

4.3.1 Competition Phase of Experimental Framework

Development of 3DV HEVC extensions is based on an experimental framework and multiview video-plus-depth (MVD) format. At the encoder side a real-world 3D scene is captured by multiple cameras, and a MVD representation is extracted from this input (Fig. 4.10). Once the **depth maps** are obtained, new views can be synthesized by interpolating the pixel values from nearby images. The *depth* of a 3D scene is expressed relative to the camera position or an origin in the 3D space. The *disparity* estimation is the correspondence between pixels in the left and right images. At the decoder side the decoder receives a coded representation (bit stream) of the data, which is then decoded and used for multiview rendering of the 3D scene.

The MPEG standardization adopted three steps of development-based formal subjective assessment of the 3D video quality:

- *Call for Evidence* (**CoE**) purpose is to explore *in house* whether the coding efficiency and 3DV functionality of the current version of HEVC standard can be further improved for MVD content.
- *Call for Proposals* (**CfP**) on 3D video coding technology is open to external parties (04/2011) with primary goal to define a 3DV data format and associated compression technology to enable the high-quality reconstruction of synthesized views for 3D displays. To evaluate the proposed technologies, formal subjective tests are performed. Results of these tests are made public (12/2011).
- *Verification tests* for 3D video coding technology include test conditions, evaluation methodology, and timeline to assess the improvement of the coding performance (10/2015) (Table 4.2).

Table 4.2 MPEG documents in competition phase of 3DV standardization

Doc. N10357 *Vision on 3D video*	Feb. 2009
Doc. N10359 *Call for 3D test material: depth maps & supplementary information*	Feb. 2009
Doc. N11631 *Report on experimental framework for 3D video coding*	Oct. 2010
Doc. N12035 *Applications and requirements on 3D video coding technology*	Mar. 2011
Doc. N12036 *Call for proposals on 3D video coding technology*	Mar. 2011
Doc. N12347 *Report of subjective test results from the call for proposals on 3D video coding tech*	Dec. 2011
Doc. N12348 *Overview of 3DV coding tools proposed in the CfP*	Dec. 2011

The *Call for Proposals on 3D video coding technology* represented the start of standardization of depth-based 3D formats, among which MVD was the first priority.

- In the CfP, two classes of test sequences (MVD format) were used as test materials. The individual sequences in each set were 8 or 10 s long.
- Two test scenarios were defined and refer to the 2-view input configuration and 3-view input configuration.
- Two test categories were defined in the CfP: AVC-compatible, and HEVC-compatible and unconstrained. For the AVC-compatible test, anchors for the objective and subjective measurements were generated using an MVC encoder (JMVC version 8.3.1) to encode the test sequences. For the HEVC compatibility test, anchors for the objective and subjective measurements were generated using an HEVC encoder (HM version 2.0) to encode the test sequences. For the AVC compatibility test, MVC was applied separately to texture data and depth data. For the HEVC compatibility test, HEVC simulcasting was used for each view of texture data and depth data. To calculate the objective rate-distortion (RD) performance and provide appropriate materials for subjective evaluation, four rate points (R1, R2, R3, and R4) were determined for each test sequence, for each test scenario, and for each test category.

Twenty-two proposals were submitted for the CfP. The submitted test materials were subjectively assessed in 12 test laboratories (18 naive viewers per test sequence) around the world. The subjective evaluations showed that, for most test sequences, the subjective quality of R3 of the best-performing proposal was better than R1 of the anchor. This suggests a significant improvement in coding efficiency compared to the anchor. In terms of objective performance, more than 25 and 55 % bitrate saving was reported by best proposals, Nokia in AVC test category and HHI in HEVC test category, respectively.

New coding tools proposed in CfP improve coding efficiency taking into account the unique *functionality* or *statistical* properties of depth data, as well as exploiting the *coherence* between texture and depth signals:

- **Texture-coding-dependent views that are independent of depth.** This involves coding the texture images of the side view. A **side view** is any view other than the first view in the coding order. The first view (also called the **base view**) is expected to be fully compatible with AVC or HEVC; the side view only uses inter-view texture information. Tools in this category include *motion parameter prediction and coding*, and *inter-view residual prediction*.

- **Texture-coding-dependent views that are dependent on depth.** This is applicable to side-view texture, in which original or reconstructed depth information is used to further exploit the correlation between texture images and associated depth maps. Tools in this category include *view synthesis prediction* for texture and *depth-assisted in-loop filtering* of texture.
- **Depth coding that is independent of texture.** Inter-view depth information or neighboring reconstructed depth values are used to compress the current macroblock in the depth map. Tools in this category include *depth intra-coding*, *synthesis-based inter-view prediction*, *inter-sample prediction*, and *in-loop filtering* for depth.
- **Depth coding that depends on texture.** Original or reconstructed texture information is used to further exploit the correlation between texture images and associated depth maps. Tools in this category include *prediction parameter coding*, *intra-sample prediction*, and *coding of quantization parameters*.
- **Encoder control optimization.** Tools in this category include *rate-distortion optimization* (RDO) techniques for depth, and texture encoding. They do not affect syntax or semantics.

4.3.2   Collaboration Phase of Experimental Framework

System structure of the best CfP proposals and coding tools from other proposals are included in the test model under consideration (TMuC) for HEVC-based 3D video coding. TMuC simulation software includes several applications and libraries for encoding, decoding, and view synthesis (Table 4.3).

The development of 3D extensions for HEVC and AVC is based on a set of core experiments (**CE**) that specifies tools under investigation and timeline of simulation and cross-check reports. Common test conditions (**CTC**) for 3DV experimentation specify test scenarios under consideration, test sequences, basic encoder configuration, and objective/subjective evaluation of visual quality (Table 4.4).

The standardization track of 3D extensions for AVC and HEVC is shown in Table 4.5.

Table 4.3 MPEG documents in the start of collaboration phase of 3DV standardization

Doc. N12350 *Test model under consideration for HEVC based 3D video coding*		Dec. 2011
Doc. N12352 *Common test conditions for 3DV experimentation* (CTC)		Dec. 2011
Doc. N12353 *Description of core experiments in AVC based 3D video coding*		Dec. 2011
Doc. N12354 *Description of core experiments in HEVC based 3D video coding*		Dec. 2011
Doc. N12434 *Standardization tracks considered in 3D video coding*		Dec. 2011

Table 4.4 Description of core experiments in AVC/HEVC 3D video coding (Dec. 2011)

CE	AVC compatible	HEVC compatible
CE1	View synthesis prediction for texture and depth	View synthesis-based prediction for texture
CE2	Depth-based prediction	View synthesis-based prediction for depth
CE3	Depth representation and coding	Motion parameter prediction and coding (independent of depth)
CE4	Depth intra-prediction without inter-component prediction	Transform coding for depth
CE5	Depth range compensation for inter/interview prediction	In-loop filtering for depth
CE6	In-loop depth resampling	Prediction parameter coding (motion data and intra-pred. mode)
CE7	RD optimization through view synthesis distortion	Coding of quantization parameters
CE8	Global depth-and-view prediction	Component extraction
CE9	Texture-based prediction for depth coding	Prediction structures for inter-view prediction
CE10	Depth in-loop filtering	Modified distortion measure for depth coding
CE11		View synthesis

Table 4.5 JCT-3V standardization track: (a) MVC+D (*multiview and depth video coding*), 3D-AVC (*multiview and depth video with enhanced non-base view coding*), (b) MV-HEVC (*multiview high-efficiency video coding*), 3D-HEVC (*3D high-efficiency video coding*)

		1. 07/2012 CD	2. 10/2012	3. 01/2013	4. 04/2013 H.264 Annex I	5. 07/2013	6. 10/2013	7. 01/2014 H.264 Annex J	8. 04/2014	9. 07/2014	10. 10/2014	11. 02/2015	12. 07/2015
Test Model	MVC+D	TM3	TM4	TM5	TM6	TM7	TM8	TM9					
	3D-AVC	3DV-ATM	3DV-ATM	3DV-ATM	3DV-ATM	3DV-ATM	3DV-ATM	3DV-ATM					
Draft Text	MVC+D	DAM2 WD4	DAM2 WD5										
	3D-AVC	WD3	WD4	WD5	WD6	WD7	WD8						
Software Draft	MVC+D				SD1	SD2	SD3			SD4	SD5		
	3D-AVC							SD2		SD3	SD4	SD5	
Conformance Draft	MVC+D		CD2	CD2	CD3	CD4	CD5	CD6					
	3D-AVC				CD1	CD2	CD3	CD4	CD5				

		1. 07/2012 CD	2. 10/2012	3. 01/2013	4. 04/2013 H.265 1E	5. 07/2013	6. 10/2013	7. 01/2014 CD	8. 04/2014	9. 07/2014	10. 10/2014 H.265 2E	11. 02/2015 CD	12. 07/2015 H.265 3E
Test Model	MV-HEVC						TM6 3D-	TM7 3D-	TM8 3D-	TM8 3D-	TM10 3D-	TM11 3D-	
	3D-HEVC	TM1 3D-HTM	TM2 3D-HTM	TM3 3D-HTM	TM4 3D-HTM	TM5 3D-HTM	HTM	HTM	HTM	HTM	HTM	HTM	
Draft Text	MV-HEVC	WD1	WD2	WD3	WD4	WD5	WD6	WD7	WD8	WD9			Verification Test Plan
	3D-HEVC					WD1	WD2	WD3	WD4	WD5	WD6	WD7	
Software Draft	MV-HEVC								SD1 HTM-MV12	SD2 HTM-MV13	SD3 HTM-MV14	SD4	
	3D-HEVC										SD1 HTM14	SD2	
Conformance Draft	MV-HEVC								CD1				
	3D-HEVC										CD2	CD3	

- MVC-compatible extension including depth MVC+D (no block-level changes to AVC/MVC syntax and decoding processes; add high-level syntax to enable efficient coding of depth data), FDAM 10/2012 (Final Draft Amendment).
- AVC-compatible extension plus depth 3D-AVC (change syntax and decoding process for non-base texture view and depth maps at block level), FDAM 07/2013.
- HEVC 3D extensions: MV-HEVC multiview extension (no change to the CU-level syntax, semantics, and decoding processes of HEVC), and 3D-HEVC (advanced multiview and 3D extension for higher compression efficiency by jointly compressing texture and depth data).

JCT-3V group developed a new data format and associated compression technology to enable the high-quality reconstruction of synthesized views for 3D displays in HEVC-based coding frameworks. As part of this work, two amendments of the HEVC standard have been developed as outlined below.

- Multiview extension (**MV-HEVC**): The main target of this extension is to enable coding multiview video sequences. Depth maps can be coupled with multiview video stream using auxiliary pictures, which are one of the features in the range extension of HEVC. There are no change to the CU-level syntax, semantics, and decoding processes of HEVC. The specification of this extension (ISO/IEC 23008-2:201x) has included in the second edition of HEVC, which has reached FDIS status in October 2014.

- 3D video extension (**3D-HEVC**): This extension has been developed that aims for higher compression efficiency by jointly compressing texture and depth data. The specification of this extension (ISO/IEC 23008-2:2013/Amd.4) has reached FDAM status in February 2015.

As the standardization of both specifications is completed, verification tests are planned to assess the improvement of the coding performance. MV-HEVC is planned to be compared with simulcast coding of HEVC as well as MVC in terms of stereo video coding. 3D-HEVC will be compared to MV-HEVC. The test conditions and evaluation procedure are based on CTC common test conditions for 3DV experimentation.

The timeline in verification test plan is as follows [Doc. N15441 July 2015]:

- Preparing viewing materials and bit streams with various bit rates (07/2015).
- Decide target bit rate for testing, perform expert viewing test with at least nine experts, and prepare the report (10/2015).

4.3.3 An Overview of 3D Video Coding Tools

4.3.3.1 MV-HEVC Coding Tools

MV-HEVC specification follows the same design principles of the MVC extension in the AVC framework. The coding schemes enable *inter-view prediction* based on **disparity-compensated prediction** (Fig. 4.11). A block-based disparity shift between the reference view and the current views is determined and used in prediction. This is similar to the motion-compensated prediction used in conventional video coding, but it is based on pictures with different viewpoints rather than pictures at different time instances. MV-HEVC extends the high-level syntax so that the appropriate signaling of view identifiers and their references is supported and defines a process by which decoded pictures of other views can be used to predict a current picture in another view.

In order to support depth map coding, MV-HEVC enables *auxiliary picture* syntax. The auxiliary picture decoding process would be the same for video or multiview video. In AVC framework, an *independent* second stream is specified for the representation of depth as well as high-level syntax signaling of the necessary information to express the interpretation of the depth data and its association with the video data. This approach does not involve macroblock-level changes to the AVC or MVC syntax, semantics, and decoding processes. The corresponding 3D video codec is referred to as **MVC + D**.

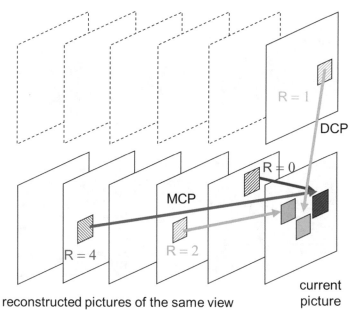

Fig. 4.11 Disparity-compensated prediction (DCP) as an alternative to motion-compensated prediction (MCP)

4.3.3.2   3D-HEVC Coding Tools for Texture

To achieve higher coding efficiency, researchers have studied and evaluated advanced coding tools that better exploit inter-view redundancy. In contrast to MV-HEVC, block-level changes to the syntax and decoding process are considered to maximize the possible coding gain. In the AVC framework, the **3D-AVC** extension supports new block-level coding tools for texture views.

Neighboring block-based disparity vector derivation (NBDV): This tool derives a disparity vector for a current block using an available disparity motion vector of spatial and temporal neighboring blocks. The derivation principle is the same in both 3D-AVC and 3D-HEVC, but the location of neighboring blocks differs slightly (Fig. 4.12). The main benefit of this technique is that disparity vectors to be used for inter-view prediction can be directly derived without additional bits and independent of an associated depth picture. Disparity information can also be derived from the decoded depth picture when camera parameters are available.

Inter-view motion prediction: The motion information between views exhibits a high degree of correlation, and inferring it from one view to another view leads to notable gains in coding efficiency because good predictions generally reduce the bit rate required to send such information. To achieve this, the disparity, such as that derived by the NBDV process, is used to establish a correspondence between the blocks in each view. The concept of inter-view motion prediction is supported

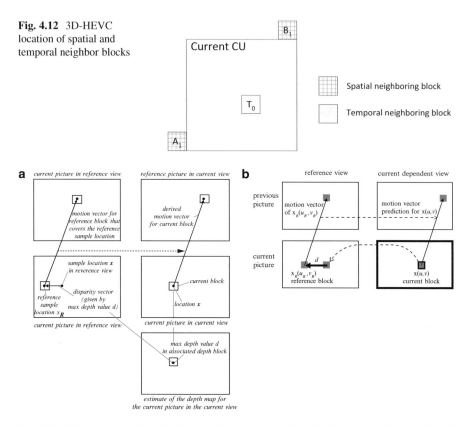

Fig. 4.12 3D-HEVC location of spatial and temporal neighbor blocks

Fig. 4.13 (**a**) Basic principle of deriving motion parameters for a block in a current picture based on motion parameters in an already coded reference view and an estimate of the depth map for the current picture. (**b**) Motion vector correspondences between a block in a current picture of a dependent view and an already coded reference view, using the disparity vector d from a depth map estimate

in both the 3D-AVC and 3D-HEVC, but the designs differ. In 3D-AVC, interview motion prediction is realized with a new prediction mode, whereas in 3D-HEVC, it is realized by leveraging the syntax and decoding processes of the merge and advance motion vector prediction (AMVP) modes that were newly introduced by the HEVC standard (Fig. 4.13).

View synthesis prediction (VSP): This tool uses the depth information to warp texture data from a reference view to the current view in order to generate a predictor for the current view. Although depth is often available with pixel-level precision, a block-based VSP scheme has been specified in both 3D-AVC and 3D-HEVC to align this type of prediction with existing modules for motion compensation. To perform VSP, the depth information of the current block is used to determine the corresponding pixels in the inter-view reference picture (Fig. 4.14). Because texture is typically coded prior to depth, the depth of the current block can be

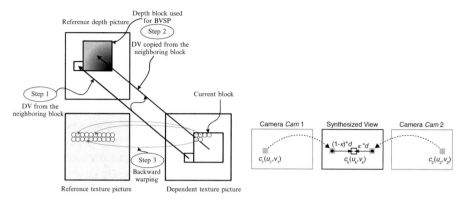

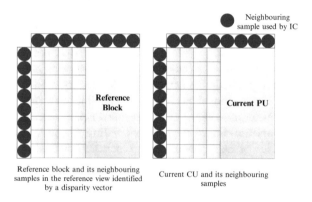

Fig. 4.14 (**a**) Illustration of the VSP scheme with the neighboring block disparity vector (DV), (**b**) view synthesis principle with horizontal disparity-based shift from original data (Cam1, Cam2) to new position in synthesized view

Fig. 4.15 Neighboring samples for the derivation of illumination compensation parameters

estimated using the NBDV process. In 3D-AVC, it is also possible to code depth prior to texture and hence obtain the depth information directly. As with inter-view motion prediction, the same VSP concept is supported in both 3D-AVC and 3D-HEVC, but the designs differ significantly. VSP is supported in 3D-AVC with a high-level *syntax flag* that determines whether the reference picture to be used for prediction is an inter-view reference picture or a synthesized reference picture as well as a low-level *syntax flag* to indicate when skip/direct mode prediction is relative to a synthesized reference picture. In 3D-HEVC, the VSP design is realized by extensions of the merge mode, whereby the disparity and inter-view reference picture corresponding to the VSP operation are added to the merge candidate list.

Illumination compensation (IC): This tool improves the coding efficiency for blocks predicted from inter-view reference pictures in case when prediction fails due to not calibrated cameras capturing the same scene or by lighting effects. This mode only applies to blocks that are predicted by an interview reference picture (Fig. 4.15).

Inter-view residual prediction: Advanced residual prediction (**ARP**) takes advantage of the correlation between the motion-compensated residual signal of two views. ARP mode only supported in 3D-HEVC increases the accuracy of the residual predictor. In ARP, the motion vector is aligned for the current block and the reference block, so the similarity between the residual predictor and the residual signal of the current block is much higher, and the remaining energy after ARP is significantly reduced. Two types of ARP designs exist: temporal ARP and inter-view ARP. In temporal ARP, the residual predictor is calculated as a difference between the reference block (Base) and its reference block (BaseRef). With inter-view ARP, an inter-view residual is calculated from the temporal reference block in a different view (BaseRef) and its inter-view reference block, hypothetically generated by the disparity (DMV) that is signaled for the current block (Fig. 4.16).

4.3.3.3   3D-HEVC Coding Tools for Depth Maps

To achieve higher compression efficiency, new coding tools have been adopted in 3D-HEVC for the coding of depth views. Depth views in 3D-AVC are coded similar to MVC+D, and no block-level changes for depth coding have been introduced.

Depth motion prediction: Similar to motion prediction in texture coding, depth motion prediction is achieved by adding new candidates into the merge candidate list. The additional candidates include interview merge candidate, subblock motion parameter inheritance candidate, and disparity-derived depth candidate.

Partition-based depth intra coding: To better represent the particular characteristics of depth, each depth block may be geometrically partitioned and more efficiently represented. In 3D-HEVC, these *nonrectangular* partitions are collectively referred to as depth modeling modes (**DMM**s), e.g., only coding the average value or predicting a planar function from already coded neighboring blocks without residual data. Two types of partitioning patterns are applied: wedglet pattern, which segments the depth block with a straight line, and contour pattern, which can support two irregular partitions.

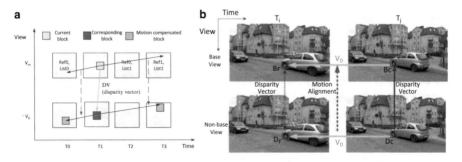

Fig. 4.16 (**a**) Relationship among current block, reference block, and motion-compensated block, (**b**) prediction structure of advanced residual prediction

Segment-wise DC coding (SDC): This coding mode enables the transform and quantization process to be skipped so that depth prediction residuals are directly coded. It also supports a depth look-up table (**DLT**) to convert the depth values to a reduced dynamic range. SDC can be applied to both intra- and inter-prediction, including the new DMM modes. When the SDC mode is applied, only one DC predictor is derived for each partition, and based on that, only one DC difference value is coded as the residual for the whole partition.

4.4   3D-HEVC Efficiency in Joint Coding-Dependent Views and Depth Data

3D-HEVC enables application that requires a high compression efficiency, such as transmitting 3D 4K content for stereoscopic as well as auto-stereoscopic multi-view displays. 3D-HEVC extension targets multiview video and depth data coding with the best coding performance. To evaluate the compression efficiency of coding tools, simulations were conducted using the reference software and experimental evaluation methodology (Fig. 4.17). In the experimental framework, multiview video and corresponding depth are provided as input, while the decoded views and additional views synthesized at selected positions are generated as output. Common test conditions (CTC) for experimentation specify basic encoder configuration, and objective/subjective evaluation of decoded/synthesized views.

New 3D-HEVC added tools for joint coding the dependent views and depth data can be clustered, according to their redundancy reduction principles: **inter-view prediction** under consideration of depth, as well as **inter-component prediction** between texture and depth pictures.

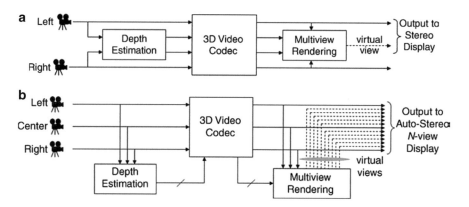

Fig. 4.17 (**a**) Advanced stereoscopic processing with two-view configuration, (**b**) auto-stereoscopic output with three-view configuration

Inter-view prediction: Similar to the compression of dependent views in MV-HEVC, the redundancy reduction across different views is one of the most important aspects for efficient coding. In addition to *disparity-compensated prediction*, 3D-HEVC uses further tools for inter-view prediction. The first tool is *view synthesis prediction*, which uses depth-based rendering to warp pixels from a reference view to a dependent view, while DCP uses one linear vector for a block. The second tool is inter-view *motion parameter prediction*. Also, motion vectors for the same content in the different views can be similar, such that they can be predicted across views, using again the depth/disparity information. Third, inter-view *residual prediction* is used. Again, also the residual data in different views is similar for a certain amount of blocks, such that prediction across views can gain coding efficiency.

Inter-component prediction: Coding tools for reducing redundancies between the video and *co-located* depth component of each view were also developed for 3D-HEVC. One *depth coding mode* **DMM4** uses texture information for depth coding. Next, the *motion parameter inheritance* checks the partitioning and motion data from the texture information, whether it can be used for efficient coding of the current depth block. Also, tools for *block partitioning prediction* can be applied, e.g., quadtree prediction, where subdivision information of the texture is used to restrict the subdivision of a co-located depth block. This assumes that the texture is finer structured than depth, such that a depth block is never subdivided further than the texture.

Encoder control: 3D-HEVC uses a **joint rate-distortion optimization (JRDO)** for the depth data. For video data, the classical rate-distortion optimization (RDO) is used, when the optimal coding mode is sought. Here, the *Lagrangian* cost function is used, a weighted sum of video rate, and video distortion in terms of mean squared error (MSE) between original and reconstructed video data. In contrast, reconstructed depth maps are only used for the synthesis of intermediate views and not directly viewed. Therefore, the coding efficiency in 3D-HEVC is improved by applying a cost function that considers the distortion in synthesized intermediate views. This **view synthesis optimization** (VSO) modified the distortion measure for the mode decision process for depth maps in a way that a weighted average of the synthesized view distortion and the depth map distortion. To obtain a measure of the synthesized view distortion, two different metrics are applied in JRDO. The distortion measurement is designed based on the fact that the same depth distortions generally cause higher synthesis errors in highly textured regions than in textureless regions.

The results obtained showed that a 3D-HEVC achieved higher coding efficiency by *optimizing* existing coding tools and *adding* new methods. In particular, an improved inter-view prediction, new methods of inter-component parameter prediction, special depth coding modes, and an encoder optimization for depth data coding towards the synthesized views were applied for optimally encoding 3DV data and synthesizing multiview video data for different 3D displays from the decoded bit stream.

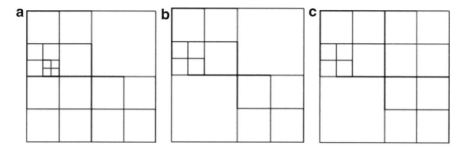

Fig. 4.18 (**a**) Example of a CTB QT partitioning for the texture, (**b**) allowed, and (**c**) disallowed collocated depth CTB QT partitioning

However, ***complexity reduction*** of an encoder is becoming a critical problem in implementations for specific application. The improvement of the 3D-HEVC coding efficiency is obtained at the expense of a computational complexity increase. The most computationally intensive parts are test all the coding modes and computation of the RD cost in recursive splitting of coding units.

Depth quadtree limitation: This tool prevents the encoder from making full investigation of every possible QT configuration for the depth coding. Based on RD optimized decisions, a given CTB is split into smaller CUs in the encoding process. A corresponding quadtree (QT) is obtained for the texture, and another one for the depth. The tool forces the encoder to limit the partitioning of the depth at the same level as the partitioning of the texture. For a given CTU, the quadtree of the depth is linked to the collocated CTB quadtree in the texture, so that a given CU of the depth cannot be split more than its collocated CU in the texture (Fig. 4.18).

Early decision algorithms: Two algorithms accelerate encoder decision by exploiting inter-view correlations in dependent texture view coding: early merge mode decision algorithm, and early CU splitting termination algorithm. Experimental results show that the proposed algorithm can achieve 47.1 % encoding timesaving with overall 0.1 % BD-rate reduction compared to 3D-HEVC test model version 7 under the common test condition (CTC). Both of the two strategies have been adopted into the 3D-HEVC reference software and enabled as a default encoding process under CTC.

4.5   Conclusion

Development of 3D video technologies is a challenging task. The current status is maturing of standardized 3D HEVC extensions and associated MVD formats. Current research issues are operational optimization of reference encoder configuration and performance improvements based on scalability extensions. New standardization activity is the next-generation video coding beyond HEVC for support of advanced 3D holoscopic representation beyond binocular cues.

Bibliography

Books

1. Assuncao P, Pinto L, Faria S (2014) Chapter 2: 3D media representation and coding. In: Kondoz A, Dagiuklas T (eds) 3D Future internet media, Springer, pp 9–38
2. Faria SMM, Debono CJ, Nunes P, Rodrigues NMM (2015) Chapter3: 3D video representation and coding. In: Kondoz A, Dagiuklas T (eds) Novel 3D media technologies. Springer, pp. 25–48
3. Kondoz A, Dagiuklas T (eds) (2016) Connected 3D media in the Internet era, Springer
4. Rao KR, Bojković Z, Milovanović D (2002) Multimedia communication systems: techniques, standards and networks. Prentice Hall PTR (Pearson Education)
5. Rao KR, Bojković Z, Milovanović D (2005) Introduction to multimedia communication: applications, middleware, networking. Wiley
6. Leonardo C (ed) (2012) The MPEG representation of digital media. Springer
7. Wien M (2015) High efficiency video coding: coding tools and specification. Springer Signals Commun Technol. (Chapter 12 extensions to HEVC, pp 291–308)
8. Möller S, Raake A (2014) Quality of experience: advanced concepts, applications and methods. Springer. (Lebreton P, Barkowsky M, Raake A, Le Callet P. Chapter 20: 3D video, pp 299–313)
9. Zhu C, Zhao Y, Yu L, Tanimoto M (2013) 3D-TV system with depth-image-based rendering: architectures, techniques and challenges, Springer. (Müller K, Merkle P, Tech G. Chapter 8: 3D video compression, pp 223–248)
10. Dufaux F, Pesquet-Popesu B, Cagnazzo M (eds) (2013) Emerging technologies for 3D video: creation, coding, transmission and rendering. Wiley. (Cagnazzo M, Pesquet-Popescu B, Dufaux F. Chapter 6: 3D video representation and formats, pp 102–120) (Vetro A, Müller K. Chapter 8: depth-based 3D video formats and coding technology, pp 139–161)
11. Zatt B, Shafique M, Bampi S, Henkel J (2013) 3D video coding for embedded devices: energy efficient algorithms and architectures. Springer
12. Ozaktas HM, Onural L (eds) (2007) Three-dimensional television: capture, transmission, and display. Springer. (Smolic A, Merkle P, Müller K, Fehn C, Kauff P, Wiegand T. Chapter 9: compression of multi-view video and associated data, pp 313–350)
13. Schreer O, Kauff P, Sikora T (eds) (2005) 3D video communication algorithms, concepts and real-time systems in human centered communication. Wiley. (Smolic A, Sikora T. Chapter 11: coding and standardization, pp 193–216)

Journals

14. (2014) Advances in 3D video processing. J Vis Comm Image Represent 25(4)
15. (2014) Special issue on QoE in 2D/3D video systems. J Vis Comm Image Represent 25(3)
16. (2013) Special section on 3D video representation, compression, and rendering. IEEE Trans Image Process 22(9)
17. (2011) Special issue 3D media and displays. Proc IEEE 99(4)
18. (2007) Special issue on 3DTV. Signal Process Image Comm. Elsevier
19. (2007) Special issue in Multiview imaging and 3DTV. IEEE Signal Process Mag
20. (2007) Special issue on Multiview video coding and 3DTV. IEEE Tran CSVT
21. (2008) Special issue on 3DTV: capture, transmission and display of 3D video. EURASIP J Adv Signal Process

B1. Introduction

22. Sullivan G, Wiegand T (2005) Video compression—from concepts to H.264/AVC standard. Proc IEEE 93(1):18–31
23. ISO/IEC JTC1/SC29/WG11 Doc. N13364 (2013) White paper on state of the art in compression and transmission of 3D video. MPEG 103. Meeting, Geneva CH

B2. Three-Dimensional Video Formats and Associated Compression Technology

24. Müller K, Merkle P, Wiegand T (2011) 3-D video representation using depth maps. Proc IEEE 99(4):643–656
25. Vetro A, Wiegand T, Sullivan G (2011) Overview of the stereo and multiview video coding extensions of the H.264/MPEG-4 AVC standard. Proc IEEE 99(4):626–642
26. Wiegand T, Sullivan GJ, Bjøntegaard G, Luthra A (2003) Overview of the H.264/AVC video coding standard. IEEE Trans Circ Syst Video Technol 13(7):560–576
27. Sullivan G, Ohm J-R, Han W-J, Wiegand T (2012) Overview of the high efficiency video coding (HEVC) standard. IEEE Trans CSVT 22(12):1649–1668
28. Ohm JR, Sullivan GJ, Schwarz H, Tan TK, Wiegand T (2012) Comparison of the coding efficiency of video coding standards—including high efficiency video coding (HEVC). IEEE Trans Circ Syst Video Technol 22(12):1669–1684
29. Bossen F, Bros B, Suhring K, Flynn D (2012) HEVC complexity and implementation analysis. IEEE Trans Circ Syst Video Technol 22(12):1685–1696
30. Sullivan GJ, Boyce J, Chen Y, Ohm J-R (2013) Standardized extensions of high efficiency video coding (HEVC). IEEE J Selected Topics Signal Process 7:1001–1016

B3. HEVC Standardization Framework

31. Chen Y, Vetro A (2014) Next-generation 3D formats with depth map support. IEEE Multimedia 21(2):90–94
32. Hannuksela MM, Rusanovskyy D, Su W, Chen L, Li R, Aflaki P, Lan D, Joachimiak M, Li H, Gabbouj M (2013) Multiview-video-plus-depth coding based on the advanced video coding standard. IEEE Trans Image Process 22(9):3449–3458
33. Müller K, Schwarz H, Marpe D, Bartnik C, Bosse S, Brust H, Hinz T, Lakshman H, Merkle P, Rhee FH, Tech G, Winken M, Wiegand T (2013) 3D high-efficiency video coding for multiview video and depth data. IEEE Trans Image Process 22(9):3366–3378
34. Domanski M, Stankiewicz O, Wegner K, Kurc M, Konieczny J, Siast J, Stankowski J, Ratajczak R, Grajek T (2013) High efficiency 3D video coding using new tools based on view synthesis. IEEE Trans Image Process 22(9):3517–3527

B4. 3D-HEVC Efficiency in Joint Coding of Dependents View and Depth Data

35. Müller K (2014) 3D extensions for high-efficiency video coding. IEEE Comsoc MMTC E-Lett 9(1)
36. Zhang N, Zhao D, Chen Y-W, Lin J-L, Gao W (2014) Fast encoder decision for texture coding in 3D-HEVC. Signal Process Image Commun 29(9):951–961
37. Zhang Y, Kwong S, Xu L, Hu S, Jiang G, Kuo C-CJ (2013) Regional bit allocation and rate distortion optimization for multiview depth video coding with view synthesis distortion model. IEEE Trans Image Process 22(9):3497–3512

Chapter 5
Depth from Defocus and Coded Apertures for 3D Scene Sensing

Erdem Sahin, Chun Wang, and Atanas Gotchev

Abstract Depth from defocus (DfD) is one of the popular passive depth sensing approaches in computer vision. Usually, it utilizes the defocus blur depth cue encoded in images, captured by a single photographic camera. Since the defocus blur is characterized by the aperture shape of the camera, it can actually be structured by inserting a coded mask in the camera aperture position. Therefore, by employing optimized masks in such coded aperture cameras, it is possible to improve the performance of DfD approaches. DfD and coded aperture approaches constitute the main theme of this chapter. Noting, however, that the stereopsis cue is actually the most widely utilized depth cue in computer vision, joint utilization of the defocus blur cue, and the stereopsis cue is another interesting topic, which is therefore also included in the discussion.

5.1 Introduction

Sensing of three-dimensional (3D) visual scenes is the first step in delivering 3D and immersive visual experience to users. In 3D scene sensing, acquiring information about scene depth is of vital importance for the successful scene reconstruction and/or ultra-realistic visualization. Depth can be sensed by specifically dedicated sensors or can be estimated from multi-perspective images. Depending on whether it requires a dedicated illumination and power or purely relies on images taken under natural light, depth sensing methods are categorized into active and passive methods. Both categories have their (application-specific) pros and cons. Passive methods are especially attractive for power-constrained (e.g., mobile) devices, since they do not require additional power. This chapter specifically addresses depth from defocus (DfD) as one of the most popular passive depth sensing approaches in computer vision.

E. Sahin (✉) • A. Gotchev
Tampere University of Technology, Tampere, Finland
e-mail: erdem.sahin@tut.fi; atanas.gotchev@tut.fi

C. Wang
Samsung R&D Institute of China, Beijing, China
e-mail: lukewang25@live.cn

© Springer Science+Business Media New York 2017
A. Kondoz, T. Dagiuklas (eds.), *Connected Media in the Future Internet Era*,
DOI 10.1007/978-1-4939-4026-4_5

DfD approaches utilize the monocular pictorial defocus blur cue. The defocus blur cue is fully characterized by the point spread function (PSF) of the camera imaging system and DfD relies on the distinguishability of PSFs corresponding to different depths. Traditionally, DfD has been using images captured by off-the-shelf cameras for which PSFs are of circular or hexagonal shape. The PSF of a camera can be, however, purposely modified by inserting a specific physical mask (code) in the camera aperture position, which then facilitates the distinction between PSFs for different depths compared to conventional apertures and ease tackling with the ill-posedness of the underlying inverse imaging problem. Due to this important role in the DfD problem, coded aperture is one of main topics of this chapter. Hence, we include an extensive discussion of it.

DfD requires only a set of images taken by a camera at fixed position, which makes it particularly attractive in single-camera applications. The stereo matching, on the other hand, is the most widely used method for obtaining scene depth information in computer vision, which requires two images captured from two different view points. Indeed, the human visual system perceives depth mainly based on the stereopsis phenomenon. Therefore, the relation between the monocular defocus blur cue and the binocular stereopsis cue, as well as their combined utilization for improved depth estimation, are interesting research topics, which are thus also discussed in this chapter.

In Sect. 5.2, modeling of the camera imaging system is presented. The properties of the defocus blur cue and a general overview of depth estimation methods based on DfD are discussed in Sect. 5.3. Having introduced the DfD concept, we present the motivation and working principles of coded aperture cameras and analyze depth estimation methods utilizing coded apertures in Sect. 5.4. In Sect. 5.5, the joint utilization of defocus blur cue and stereopsis cue is explored for improved depth estimation. In particular, several stereo camera systems equipped with coded apertures are presented. The content presented in this chapter is based on the works [29, 30].

5.2   Camera Imaging System

In this section, we discuss how an image of a 3D scene is captured by a camera imaging system, since understanding the imaging process is important for establishing the DfD problem. During the discussion below, we consider a simplified yet representative case where both the camera and scene are still, and the camera lens is assumed to be free of aberrations. Those mild idealizations help us to avoid collateral problems such as motion blur, which would unnecessarily complicate the presentation at this stage, and focus on the main DfD problem.

The imaging process constitutes three main components: a 3D scene to be imaged as the signal source, a camera imaging system that captures and processes the signal, and the captured image as the result of the process. The 3D scene can be viewed as a cloud of self-luminous point light sources on objects surfaces. For each point light

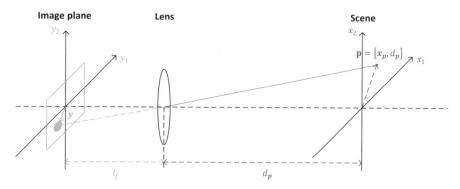

Fig. 5.1 The image formation process and the coordinate system, where the lens center is taken as the origin

source, its position in the scene space can be traced by a vector $p \in \mathbb{R}^3$. That is, p traces the surface of objects in the 3D scene. By setting the optical center of the camera imaging system (i.e., the center of the lens in the thin-lens model) at origin, this vector p can be further separated into two parts, i.e., $p = [x_p, d_p]$, where $d_p \in \mathbb{R}$ denotes the depth of the point source and $x_p = [x_1, x_2] \in \mathbb{R}^2$ denotes the position of the point source on the plane at depth d_p parallel to the principal plane of the lens (and image plane). An example point light source is shown in Fig. 5.1.

Under the Lambertian assumption, the appearance of a 3D scene can be considered as an unknown intensity distribution over the space [3], which is denoted by $f^0(p)$ and called scene intensity function. Considering the simplest case, where the scene has only a single unit intensity point source at p, at depth d_p, its image appearing in the image plane is known as the PSF, denoted by $k^{c,d_p}(y,p)$, as shown in Fig. 5.1. Here, c is used to represent the camera parameters on which the PSF is dependent. According to the thin-lens model, the camera setting parameters are mainly the aperture shape, the focal length, and the focused distance. Due to the limited physical size and viewing angle of a lens as well as the complex structure of the 3D scene, usually there exists occlusions between different objects in the 3D scene and/or self-occlusions between different parts of the same object. Consequently, not all point light sources of the scene are equally visible by the lens. Thus, considering the possible occlusions, it is reasonable to adopt "effective" aperture shape concept. The effective aperture shape can be defined as the binary masked version of the camera aperture, where the mask is one (open) at locations where the point source is visible and it is zero (closed) if the point is occluded. The camera setting $c(p)$ is, therefore, assumed to be dependent on point source (location). Hence, the PSF is denoted as $k^{c(p),d_p}(y,p)$.

Since the camera imaging system is linear, the image of an arbitrary scene can be obtained by super-positioning images of all point light sources of the scene. Thus, in the image plane, the noise-free image $g^0(y)$ of a scene with intensity function $f^0(p)$ can be obtained as

$$g^0(\boldsymbol{y}) = \int_{\mathbb{R}^3} k^{c(\boldsymbol{p}),d_p}(\boldsymbol{y},\boldsymbol{p}) f^0(\boldsymbol{p}) \mathrm{d}\boldsymbol{p}. \tag{5.1}$$

Due to the blurring effect of PSF, g^0 is also known as a blurred image of the corresponding scene intensity function f^0 [3]. Similar to the scene intensity function $f^0(\boldsymbol{p})$, $g^0(\boldsymbol{y})$ can be viewed as an intensity distribution on the image plane, which is recorded by the sensor.

Taking into account the noise and modeling the image sensor, in $\Gamma \in \mathbb{R}^2$, as a 2D rectangular lattice of M pixels, the captured noisy image $\boldsymbol{g}_M$ can be written as

$$g_M[\boldsymbol{m}] = \int_\Gamma r_m(\boldsymbol{y}) g^0(\boldsymbol{y}) \, \mathrm{d}\boldsymbol{y} + \omega_M[\boldsymbol{m}], \tag{5.2}$$

where the detector's response r_m is a weight kernel at the discrete image index $\boldsymbol{m} = [m_1, m_2]$, and ω_M represents the sensor noise on the discrete image plane. Although the sensor noise might be intensity-dependent in practice [15], here the sensor noise is assumed to be additive and is an independent and identically distributed (i.i.d.) random variable, which follows, for example, a Gaussian distribution.

Considering the discrete case, for a cloud of N point light sources, the scene intensity function can be represented as $f_N^0[n]$ on a discrete set of positions $\{\boldsymbol{p}_n\}$, where $n = 1, 2, \ldots, N$. The discrete-to-discrete imaging process can then be written as

$$g_M[\boldsymbol{m}] = \sum_{n=1}^N h^{C_N[n], D_N[n]}[\boldsymbol{m}, n] f_N^0[n] + \omega_M[\boldsymbol{m}], \tag{5.3}$$

where $h^{C_N[n], D_N[n]}[\boldsymbol{m}, n] = \int_\Gamma r_m(\boldsymbol{y}) k^{c(\boldsymbol{p}_n), d_{p_n}}(\boldsymbol{y}, \boldsymbol{p}_n) \, \mathrm{d}\boldsymbol{y}$ denotes the discrete PSF and C_N and D_N are vectors representing effective camera settings and depths of all N point light sources, respectively. Equation (5.3) can be equivalently presented in a matrix-vector multiplication form as suggested in [16]. As mentioned above, $\boldsymbol{g}_M$ and ω_M are M-dimensional noisy image vector and noise vector, respectively; f_N^0 is a scene intensity vector of dimension N. Those three vectors are linked by the camera system matrix H_{C_N, D_N} of size $M \times N$, whose nth column is the discrete PSF $\boldsymbol{h}^{C_N[n], D_N[n]}$ corresponding to the nth point light source of unit intensity. Thus, we have

$$g_M = H_{C_N, D_N} f_N^0 + \omega_M. \tag{5.4}$$

Now, let us consider a special case where the scene contains merely a fronto-parallel plane. In this case, all points in the scene space share the same depth d and both self-occlusions and occlusions are inherently avoided, which leads to a space-invariant PSF $k^{c,d}$. Thus Eq. (5.2) can be simplified as

$$g(y) = g^0(y) + \omega(y)$$

$$= \int_{\mathbb{R}^2} k^{c,d}\left(y, x_p\right) f^0(x_p) dx_p + \omega(y)$$

$$= \frac{1}{\alpha^2} \int_{\mathbb{R}^2} k^{c,d}\left(y, \frac{\tilde{x}_p}{\alpha}\right) f^0\left(\frac{\tilde{x}_p}{\alpha}\right) d\tilde{x}_p + \omega(y), \tag{5.5}$$

where $\tilde{x}_p = \alpha x_p$ with $\alpha = \dfrac{-l_f}{d}$ representing the lens magnification, and l_f is the distance between the lens and the image plane as shown in Fig. 5.1. Let $\tilde{k}^{c,d}\left(y, \tilde{x}_p\right) \triangleq k^{c,d}\left(y, \dfrac{\tilde{x}_p}{\alpha}\right)$ and $\tilde{f}^0(\tilde{x}_p) \triangleq f^0\left(\dfrac{\tilde{x}_p}{\alpha}\right)$, then Eq. (5.5) becomes

$$g(y) = \frac{1}{\alpha^2} \int_{\mathbb{R}^2} \tilde{k}^{c,d}\left(y, \tilde{x}_p\right) \tilde{f}^0(\tilde{x}_p) d\tilde{x}_p + \omega(y)$$

$$= \frac{1}{\alpha^2} \int_{\mathbb{R}^2} \tilde{k}^{c,d}\left(y - \tilde{x}_p\right) \tilde{f}^0(\tilde{x}_p) d\tilde{x}_p + \omega(y). \tag{5.6}$$

Thus, Eq. (5.6) can be simply given as

$$g = \frac{1}{\alpha^2} \tilde{k}^{c,d} \otimes \tilde{f}^0 + \omega, \tag{5.7}$$

where $\otimes$ denotes convolution. Similarly, in the discrete case, Eq. (5.4) becomes

$$g_M = \frac{1}{\alpha^2} \tilde{H}_{c,d} \tilde{f}_N^0 + \omega_M, \tag{5.8}$$

where $\tilde{H}_{c,d}$ is now a convolution matrix. Although it is often unrealistic to treat the whole camera imaging system as space invariant, usually this assumption is locally valid, since in most of the cases the structure of a 3D scene can be treated as piecewise planar. Indeed, the locally space-invariant PSF assumption is utilized in most DfD approaches, as will be discussed in Sect. 5.3. Please note that in the rest of the chapter, the scene intensity function f^0 is assumed to be scaled such that $|\alpha| = 1$, and thus the tilde sign is removed for simplicity.

5.2.1 Aperture Superposition Principle

As presented above, the PSF characterizes the camera imaging system. It is, therefore, the key element of DfD approaches, which will be addressed in Sect. 5.3. Aperture superposition principle is one way of modeling the PSF. Lanman et al. [11]

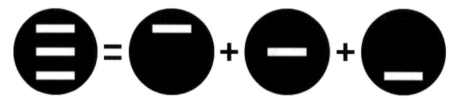

Fig. 5.2 An example of aperture superposition

showed that an aperture can be viewed as a combination of elementary apertures. In other words, the image captured with the whole aperture can be approximated by super-positioning images captured with those elementary apertures. An example of this approach is shown in Fig. 5.2. Mathematically, the aperture superposition principle can be expressed as

$$g_M = \sum_i \alpha_i g_{M_i},\tag{5.9}$$

where α_i is the transmission coefficient of the ith elementary aperture and g_{M_i} is the image captured with this elementary aperture.

The aperture superposition principle is useful in handling the effective aperture concept mentioned in Sect. 5.2. It is, however, more importantly, utilized in coded aperture modeling, which is a topic of vital significance for the DfD problem, as will be extensively discussed in Sect. 5.4. Besides its simplicity, the aperture superposition principle might not be accurate in modeling the PSF depending on the choice of elementary aperture size. Since it is a ray optics treatment of the imaging system, the elementary apertures should be wide enough so that the ray optics assumption is valid. Otherwise, diffraction effects will be visible, and wave optics treatment will be necessary.

5.2.2 Wave Optics-Based PSF Modeling

The PSF can be modeled more accurately by employing wave optics and thus taking into account diffraction effects, that could be visible, for example, if a sufficiently fine-structured mask is inserted at camera aperture position, as illustrated in Fig. 5.3.

Considering temporally coherent (monochromatic) as well as spatially coherent illumination, let us assume that we have an "equivalent thin-lens model" for our camera imaging system where we can also assume that the place of aperture and lens plane is coincident [2]. We also assume that the lens is aberration-free. Following the setup illustrated in Fig. 5.3 and utilizing the Fresnel diffraction model (under paraxial approximation) [9] we can write

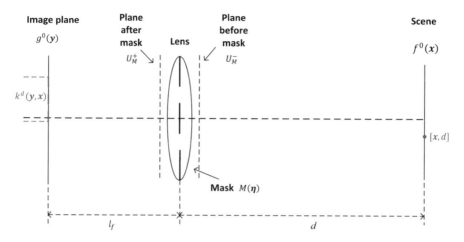

Fig. 5.3 Parameterization of the imaging system for wave optics-based PSF modeling

$$U_M^-(\eta) = \mathcal{F}\mathcal{P}_d\{f^0(x)\},\tag{5.10}$$

$$U_M^+(\eta) = U_M^-(\eta)\,M(\eta)\,P(\eta)\exp\left\{-j\frac{\pi}{\lambda f}\left(\eta_1^2+\eta_2^2\right)\right\},\tag{5.11}$$

$$g^0(y) = \mathcal{F}\mathcal{P}_{l_f}\{U_M^+(\eta)\},\tag{5.12}$$

where λ is the wavelength of monochromatic light, f is the focal length of the lens, and $M(\eta)$ is the mask function; $U_M^-(\eta)$ and $U_M^+(\eta)$ are the wave fields just before and just after the lens plane, respectively; $P(\eta)$ is the lens pupil function defined as

$$P(\eta) = \begin{cases} 1, & \text{inside the lens aperture;} \\ 0, & \text{otherwise;} \end{cases}\tag{5.13}$$

and $\mathcal{F}\mathcal{P}_z\{U(x)\}$ denotes the Fresnel propagation of $U(x)$ by distance z, which is defined as

$$\mathcal{F}\mathcal{P}_z\{U(x)\} \triangleq U_z(y)$$

$$= \frac{\exp\left(j\dfrac{2\pi z}{\lambda}\right)}{j\lambda z}\iint\limits_{\mathbb{R}^2} U(x)\exp\left\{j\frac{\pi}{\lambda z}[(y_1-x_1)^2+(y_2-x_2)^2]\right\}dx_1 dx_2.$$

$$\tag{5.14}$$

Hence, for a point light source at $[x, d]$, we have the coherent impulse response k_{coh} as [9]

$$k_{\text{coh}}^d(y, x) = A(y, x) \iint_{\mathbb{R}^2} \tilde{P}(\eta) \exp\left\{j\frac{\pi z_d(\eta_1^2 + \eta_2^2)}{\lambda}\right\}$$

$$\exp\left\{-j\frac{2\pi}{\lambda l_f}[(y_1 - \alpha x_1)\,\eta_1 + (y_2 - \alpha x_2)\,\eta_2]\right\}\,d\eta_1 d\eta_2,$$

(5.15)

where

$$\tilde{P}(\eta) = M(\eta)\,P(\eta),$$

(5.16)

$$z_d = \frac{1}{d} + \frac{1}{l_f} - \frac{1}{f},$$

(5.17)

$$\alpha = -\frac{l_f}{d},$$

(5.18)

$$A(y, x) = \frac{\exp\left\{j\frac{2\pi}{\lambda}(d + l_f)\right\}}{\lambda^2 d l_f} \exp\left\{j\frac{\pi}{\lambda l_f}(y_1^2 + y_2^2)\right\} \exp\left\{j\frac{\pi}{\lambda d}(x_1^2 + x_2^2)\right\}.$$

(5.19)

Note that the imaging system is shift-invariant for the scaled scene coordinates $(\tilde{x}_1, \tilde{x}_2) = (\alpha x_1, \alpha x_2)$, i.e., k_{coh} is a function of $(y_1 - \tilde{x}_1, y_2 - \tilde{x}_2)$.

If the illumination is perfectly spatially incoherent, but still monochromatic, the imaging system behaves linearly in intensity rather than amplitude, and in this case, the incoherent impulse responses k_{inc} is given in terms of the coherent PSF as [30]

$$k_{\text{inc}}^d(y_1 - \tilde{x}_1, y_2 - \tilde{x}_2) = |k_{\text{coh}}^d(y_1 - \tilde{x}_1, y_2 - \tilde{x}_2)|^2$$

$$= \left|\frac{1}{\lambda^2 d l_f} \iint_{\mathbb{R}^2} \tilde{P}(\eta) \exp\left\{j\frac{\pi z_d(\eta_1^2 + \eta_2^2)}{\lambda}\right\} \exp\left\{-j\frac{2\pi}{\lambda l_f}[(y_1 - \tilde{x}_1)\,\eta_1 + (y_2 - \tilde{x}_2)\,\eta_2]\right\}\,d\eta_1 d\eta_2\right|^2.$$

(5.20)

The PSF k_{inc}^d obtained for the monochromatic case can be further generalized to polychromatic illumination, by taking into account all desired spectral components Λ. If the imaging for the monochromatic and spatially incoherent case is given as

$$g^0(y) = \iint_{\mathbb{R}^2} f^0(\tilde{x}; \lambda_0)\,k_{\text{inc}}^d(y_1 - \tilde{x}_1, y_2 - \tilde{x}_2, \lambda_0)\,d\tilde{x}_1 d\tilde{x}_2,$$

(5.21)

then for the polychromatic case, it can be written as

$$g^0(y) = \iint_{\mathbb{R}^2}\int_{\Lambda} f^0(\tilde{x}; \lambda)\,k_{\text{inc}}^d(y_1 - \tilde{x}_1, y_2 - \tilde{x}_2, \lambda)\,d\lambda d\tilde{x}_1 d\tilde{x}_2.$$

(5.22)

For the discrete case, similar analysis described previously in Sect. 5.2 is also applicable here.

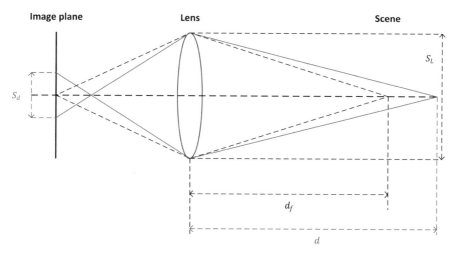

Image plane   Lens   Scene

Fig. 5.4 Illustration of the defocus blur cue for the thin-lens camera model

5.3 Depth from Defocus

When a 3D scene is sensed and recorded by optical systems, only a limited part of the scene is kept in focus; objects out of this in-focus region are inevitably captured blurred. That is, most optical systems have limited depth of field. Usually, this phenomenon is disfavored and is treated as a drawback of the optical system, which has to be compensated. Consecutively, various methods for image deblurring have been developed. However, since the degree of defocus blur reflects the relative depth of the object with respect to focus distance, it actually exhibits a depth cue [19]. As illustrated in Fig. 5.4, also revealed by the image formation model given by Eq. (5.4), the defocus blur is characterized by the PSF of the camera imaging system. The scale S_d of defocus blur, on the other hand, is related to depth d through the thin-lens imaging equation as

$$S_d = S_L \left(\left| \frac{fd_f}{(d_f - f)d} - \frac{d_f}{d_f - f} + 1 \right| \right) \tag{5.23}$$

$$\approx \frac{fS_L}{d}, \tag{5.24}$$

where f is the focal length of the lens, S_L is the width of lens aperture, and d_f is the focused distance, as denoted in Fig. 5.4. Please notice that when $d_f \gg f$, the depth-defocus blur degree relation can be assumed to be independent to the focused distance, as given by Eq. (5.24).

The DfD problem can be expressed as estimation of the correct (scale of) PSF at a given image location, given N images $\{g_{M_n}\}$ captured from a single view with known camera settings, where both the camera and the 3D scene are assumed to be fixed. According to the model given by Eq. (5.4), the depth information D_N is independent to the scene intensity function f_N^0, which is also unknown. Estimating f_N^0, on the other hand, is known as the image restoration, which might be a part of DfD depending on the algorithm as will be discussed in this Sect. 5.3.1.

We choose to analyze the DfD problem using the Bayesian inference, since it provides a sufficiently general framework. Under the Bayesian framework, the depth map D_M and the all-in-focus image f_M as well as the captured image g_M and noise ω_M are all viewed as random vectors (or images) with probability distributions, denoted by $p(D_M), p(f_M), p(g_M)$, and $p_{\omega_M}(\omega_M)$, respectively. Particularly, the joint distribution of D_M, f_M, and g_M, denoted by $p(D_M, f_M, g_M)$, gives a complete probabilistic description of the whole system. Please note that we set the resolutions of the desired all-in-focus image and its corresponding depth map to the resolution of the observed images. The case where $N > M$ is addressed in the super-resolution problem, which is out of context of this chapter.

By using Bayes' rule, we get

$$p(D_M, f_M, g_M) = p(D_M, f_M | g_M) p(g_M)$$
$$= p(g_M | D_M, f_M) p(D_M) p(f_M). \qquad (5.25)$$

When captured images, which are observations g_M, are taken into account, we have

$$p\left(D_M, f_M | g_{M_{1,...,N}}\right) \propto p\left(g_{M_{1,...,N}} | D_M, f_M\right) p(D_M) p(f_M), \qquad (5.26)$$

where $p(D_M)$ and $p(f_M)$ are prior distributions of D_M and f_M, respectively, that contain a priori information and thus introduce additional constraints to the system. $p\left(g_{M_{1,...,N}} | D_M, f_M\right)$ is the likelihood measuring the probability that images are generated by the scene information D_M and f_M. Finally, $p\left(D_M, f_M | g_{M_{1,...,N}}\right)$ is known as the joint posterior distribution of D_M and f_M, and it is the distribution of interest since the pair $\{D_M^*, f_M^*\}$ maximizing this distribution is considered as the best solution of the problem. That is, the problem is presented as a *maximum a posteriori* (MAP) probability estimation,

$$D_M^*, f_M^* = \underset{D_M, f_M}{\arg\max} \, p\left(D_M, f_M | g_{M_{1,...,N}}\right)$$

$$= \underset{D_M, f_M}{\arg\max} \, p\left(g_{M_{1,...,N}} | D_M, f_M\right) p(D_M) p(f_M)$$

$$= \underset{D_M, f_M}{\arg\max} \, \prod_{n=1}^{N} \{p(g_{M_n} | D_M, f_M)\} p(D_M) p(f_M). \qquad (5.27)$$

The way how f_M is treated leads to two categories of solution strategies for the DfD problem. In the first one, depth estimation and image restoration are solved simultaneously. That is especially useful for applications in which both the resulting depth map and the all-in-focus image are demanded. In the other case, the image restoration is bypassed and the only focus is on the depth estimation.

5.3.1 Solving Strategies: Restoration Based

In this section, the methods that follow the restoration-based strategy are introduced. Recalling Sect. 5.2, the captured image can be viewed as depth-dependent blurring of a hidden all-in-focus image by the PSFs. Thus, obtaining accurate all-in-focus image is a key step, if one wants to utilize it to reach depth information via estimation of correct PSF. The locally space-invariant PSF assumption introduced in Sect. 5.2 is mostly valid and, therefore, it is employed to simplify the relation between the captured image and all-in-focus image.

By substituting Eq. (5.7) into Eq. (5.27), the problem can be expressed patch-wisely as follows:

$$
D_M^*[l], f_M^*[l] = \underset{d, f_M}{\arg\max} \sum_L \prod_{n=1}^{N} \left\{ p_{\omega_M} \left(g_{M_n} - h^{c_n, d} \otimes f_M \right) \right\} p(f_M), \tag{5.28}
$$

where L represents the image patch around the current pixel l and $\sum_L$ denotes summation of pixel values within that patch. Please notice that $p(D_M)$ is dropped, since it is a constant within the patch L. Equation (5.28) can be separated into two parts and solved consecutively. The first step is image restoration, which aims to restore the all-in-focus image patch given PSFs for fixed depths. This step is repeated on a set of finite depths $\mathcal{K} = \{d_k\}$, and each restored image is denoted as $\hat{f}_M^k$.

The second step is image reproduction. Firstly, those restored images $\hat{f}_M^k$ and corresponding candidate PSFs $\{h^{c_n, d_k}\}$ are put in pairs. For example, the kth pair is $\left\{ \hat{f}_M^k, \{h^{c_n, d_k}\} \right\}$. Those pairs are then used to estimate captured images, where the best pair is expected to produce images mostly similar to captured ones. This selection step is actually a maximum likelihood estimation (MLE) given as

$$
D_M^*[l], f_M^*[l] = \underset{d_k, \hat{f}_M^k}{\arg\max} \sum_L p \left(g_{M_{1...N}} | h^{c_1, d_k}, \ldots, h^{c_N, d_k}, \hat{f}_M^k \right)
$$

$$
= \underset{d_k, \hat{f}_M^k}{\arg\min} \sum_L \sum_{n=1}^{N} \left(\left\| g_{M_n} - h^{c_n, d_k} \otimes \hat{f}_M^k \right\|_2^2 \right). \tag{5.29}
$$

The procedure is repeated for all pixels to acquire a dense depth map D_M^* and a restored all-in-focus image f_M^*.

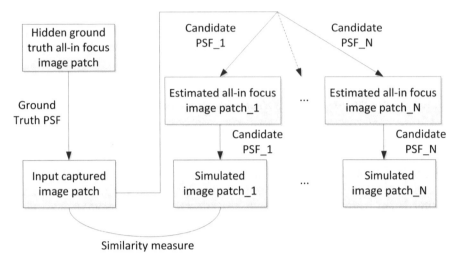

Fig. 5.5 The principle of restoration-based strategy

The restoration-based strategy is illustrated in Fig. 5.5, where a captured image patch is assumed to be generated by convolution of the ground truth all-in-focus image patch with the ground truth PSF. Ideally, when the correct PSF is selected, the estimated all-in-focus image patch and the PSF should be able to produce a simulated image patch that is similar to the captured one. For an incorrect PSF, errors will be introduced in both the image restoration step and the image reproduction step, and thus the simulated image patch will deviate from the captured one. It is obvious that the quality of the restored image is of vital importance; a failure in the image restoration will produce erroneous results.

Methods following the restoration-based strategy have been successfully applied by Levin et al. [12] in the spatial domain, for a single captured image, and by Zhou et al. [33] in the frequency domain, for a pair of captured images.

5.3.2 Solving Strategies: Restoration Free

In this section, the restoration free strategy is introduced. Methods following this strategy can be further divided into two categories as subspace-based identification and feature-based identification.

5.3.2.1 Subspace-Based Identification

The idea of subspace-based identification is to utilize the spectral supports $\mathscr{B}$ of PSFs. In the case of noise-free imaging, the captured images degraded by the same PSF share a common frequency support and thus form a subspace of the image space. Thus, depth estimation can be done by identifying the correct subspace. In most of the cases, however, it is unrealistic to ignore noise, which, mostly, randomly change the power spectrum of an image and thus make the image deviated from its subspace. Especially, its influence is heavy for spectral regions which includes zero-crossings or negligible amplitudes, since those regions are actually utilized for depth estimation purpose, as will be discussed in Sect. 5.4.1. In order to eliminate such effects of noise, a band-limiting operator $\mathscr{P}_{\mathscr{B}}$ projecting the image onto the frequency support $\mathscr{B}$ of a PSF is needed [3].

For the single image case, employing the locally space-invariant PSF assumption introduced in Sect. 5.2, the problem stated in Eq. (5.27) can be expressed patch-wisely as

$$D_M^*[l] = \arg\max_d p\left(g_L | h^{c,d}\right)$$

$$= \arg\max_d p\left(g_L | \mathscr{P}_{\mathscr{B}}^{c,d}\right). \tag{5.30}$$

Please notice that each time only a single image patch g_L is utilized. Similar to the procedure presented in Sect. 5.3.1, depth estimation is solved pixel-wise with PSFs pre-sampled at a finite set of depths $\mathscr{K}$, and it is done in two steps. The first step is to construct the band-limiting operator for all depths $d_k \in \mathscr{K}$. Then in the second step, those operators are applied to the captured image. For each image patch, the operator leading to the minimum changes indicates the most suitable subspace for this image patch, and hence the corresponding depth is proposed as a depth estimate. That is,

$$D_M^*[l] = \arg\min_d \left(\left\|g_L - \mathscr{P}_{\mathscr{B}_k}^{c,d_k} g_L\right\|_2^2\right) \tag{5.31}$$

The procedure is repeated for all pixels to acquire a dense depth map D_M^*. Obviously, the key point of subspace-based identification strategy is to determine the projection operators. Methods following this strategy has been well studied by Lin et al. [14] in the frequency domain, for the single image case. Favaro and Soatto [7], on the other hand, apply the strategy in the spatial domain to a pair of images, where instead of projecting image patches onto the subspace defined by a PSF, they project image patches onto the orthogonal subspaces, to acquire projection errors directly. This procedure was then applied by Martinello and Favaro [16] to single image case.

5.3.2.2   Feature-Based Identification

The feature-based identification aims to distinguish images blurred by different PSFs via feature vectors. The idea is quite similar to the one introduced in Sect. 5.3.2.1. However, instead of using the whole frequency supports to distinguish PSFs, only selected frequency components, termed features, are used.

For the single image case, employing the locally space-invariant PSF assumption introduced in Sect. 5.2, the problem stated in Eq. (5.27) can be expressed patch-wisely as follows:

$$
\begin{aligned}
\boldsymbol{D}_M^*[l] &= \arg\max_d \, p\left(\boldsymbol{g}_L | \boldsymbol{h}^{c,d}\right) \\
&= \arg\max_d \, p\left(\mathscr{F}\boldsymbol{g}_L | \mathscr{F}\boldsymbol{h}^{c,d}\right),
\end{aligned}
\tag{5.32}
$$

where $\mathscr{F}\boldsymbol{g}$ represents the feature vector extracted from the image $\boldsymbol{g}$ by the application of the filter bank $\mathscr{F}$. Depth estimation is solved pixel-wise with PSFs pre-sampled at a finite set of depths $\mathscr{K}$, and it is done in two steps. The first step is to design a filter bank for all depths $d_k \in \mathscr{K}$. Then in the second step, those filter banks are applied to both the captured image and the corresponding PSFs. For each image patch, the filter bank maximizing the likelihood function given in Eq. (5.32) indicates the most suitable PSF and hence the estimated depth. A dense depth map $\boldsymbol{D}_M^*$ is obtained by repeating the procedure for all pixels. As we shall discuss in Sect. 5.4, the features extracted by applying filter banks are usually the zero-crossings in the spectrums of PSFs. Thus, if the image patch has strong spectral content at (or around) the zero-crossings, it is unlikely that it is generated by this PSF. Methods of this category have been studied by Zhu et al. [34] and Burge and Geisler [4] for the single image case.

5.4 Coded Aperture

In this section, coded aperture is introduced as a technique for improving the performance of DfD. First, the motivation of inserting a coded mask in the aperture position is given from DfD point of view and then the principles of coded aperture as well as review of two mask pattern optimization strategies are presented.

5.4.1 PSF Modification

As discussed in the previous section, the PSF plays a key role in the success of DfD, since the defocus blur cue is characterized by the PSF of the imaging system. Traditional DfD approaches use images captured by off-the-shelf cameras for which PSFs are of circular or hexagonal shape. However, the PSF derived from a conventional aperture works like a low-pass filter, which attenuates high frequency components heavily, so that the comparison can only be done with remained low frequency components. In addition, the lack of high frequency components makes image restoration, if required, considerably difficult. Furthermore, a conventional aperture has a symmetrical shape, which leads to similar blurring effect on different sides of the focused distance. Hence, when only a single image is available, distinguishing the correct side of the focused distance is challenging, which is known as the sign problem [24]. Those issues not only limit the success of DfD, but also make it prone to noise. The insightful statement of Hiura and Matsuyama then naturally arises: the essence of DfD is to design desired blurring effect [10]. In this respect, two major questions need to be answered: what is the desired blurring effect and how to achieve it?

Regarding properties of blurring effect in the single image DfD case, Dowski and Cathey [5, 6] pointed out a necessary condition that the spectrum of the PSF must have regions of zero-crossings. Thus, according to this condition the depth information lies in the zero-crossings of the spectrums of the PSFs such that the zero-crossings will change position by the scale change of the PSF when traced over different depths. They also experimentally observed that it is better to have PSFs for which the zero-crossings appear periodically and the spectrum amplitudes of PSFs are as high as possible at non-zero parts, especially in the vicinity of zero-crossings [5]. By taking into account the noise effect, Levin et al. [12] further pointed out that for the depth estimation purpose, it is better to have the zero-crossings at low frequencies. This is because most of images' energies are concentrated at low frequencies, so, in that case, zero-crossings at these regions will be clearly distinguishable and thus be more robust to noise. On the other hand, Dowski and Cathey [6] noted that for image restoration, the spectrum of the PSF should be modified in such a way that it has a wide passband without zero-crossings not to lose information. That means, depth estimation and image restoration set different, even contradictory, preferences on the (spectral) properties of PSF, which is problematic especially for restoration-based DfD approaches discussed in Sect. 5.3.1. This problem will be further addressed in this section and Sect. 5.5.4.

The PSF of the imaging system can be modified by inserting a specific physical mask (code) in the camera aperture position (such cameras are referred to as coded aperture camera). In the following sections, two general mask optimization frameworks are presented that are used to have the desired blurring effect and thus achieve improved DfD performance.

5.4.1.1 Brute Force Search

Assuming that the coded mask M is of square shape, it can be viewed as formed by $n \times n$ small squares, which are considered as elementary masks $\{M_i\}$ [11]. By denoting the transmission efficiency of ith elementary mask as α_i, the aperture mask can be represented as

$$M = \sum_{i=1}^{n^2} \alpha_i M_i, \tag{5.33}$$

where n and the value range of α control the fineness of masks.

In brute force search method, optimized mask patterns are selected in two steps, namely generation and testing. The generation step aims to generate a large set of masks as candidates according to Eq. (5.33). There are, however, few practical issues need to be considered regarding this approach. Increasing n leads to finer masks, hence suboptimal masks can be improved to have optimal ones. Decreasing the size of each elementary aperture, on the other hand, may cause heavy diffraction effects. Therefore, n should be chosen properly. The other issue is the choice of available values for α_i. Generally, as a coefficient representing transmission efficiency, α_i could take any value between 0 and 1, from completely blocking to completely transmitting. However, this makes the searching space infinitely large and intractable. Therefore, when the brute force strategy is employed, generally the value of α_i is restricted to be either 0 or 1, meaning either a close or an open elementary mask, which leads to a binary mask. There are 2^{n^2} candidates for the single binary mask case, while for the case of a pair of binary masks, the number becomes 4^{n^2}. Additional constraints may be applied to eliminate undesired candidates. For example, from manufacturing point of view, the pattern should lead to a complete mask without unconnected floating parts [12]; and the pattern should have a sufficiently large optical efficiency [17].

In the testing stage, all valid candidates are evaluated to find the optimal mask. In order to do evaluation, proper criteria should be defined in accordance with the task and solving strategy. For example, to optimize a single mask for depth estimation purpose, Levin et al. [12] proposed a depth discrimination criterion. For the optimal mask c^* which has the best depth discrimination capacity, any pair of PSFs derived from it should be easily distinguishable according to some criterion, i.e., the likelihood distributions $p(\mathbf{g}_M|\mathbf{h}^{c^*,d_{k_i}})$ and $p(\mathbf{g}_M|\mathbf{h}^{c^*,d_{k_j}})$ should be as different as possible. This difference is, for example, measured using the Kullback–Leibler(KL) divergence as [12]

$$
\begin{aligned}
D_{\mathrm{KL}} &\left\{ p\left(\mathbf{g}_M|\mathbf{h}^{c^*,d_{k_i}}\right), p\left(\mathbf{g}_M|\mathbf{h}^{c^*,d_{k_j}}\right) \right\} \\
&= \int_{\mathbf{g}_M} p\left(\mathbf{g}_M|\mathbf{h}^{c^*,d_{k_i}}\right) \left\{ log\left[p\left(\mathbf{g}_M|\mathbf{h}^{c^*,d_{k_i}}\right) \right] - log\left[p\left(\mathbf{g}_M|\mathbf{h}^{c^*,d_{k_j}}\right) \right] \right\} d\mu\left(\mathbf{g}_M\right).
\end{aligned}
$$

$$(5.34)$$

Thus, for each mask candidate, PSFs are derived corresponding to a finite set of depths $\mathcal{K}$. Then, Eq. (5.34) is calculated for all pairs of PSFs to measure the depth discrimination capability. Finally, the mask candidate with the best depth discrimination capacity is chosen as the optimized single mask for DfD purpose. An example of optimized mask for $n = 13$, which we call Levin's mask, is shown in Fig. 5.6a.

Similar procedures, on the other hand, have also been employed to design single mask for image restoration purpose [17, 31]. Recalling the contradictory requirements set by the depth estimation and image restoration, which actually create problem for restoration-based DfD approaches, the mask pair reported by

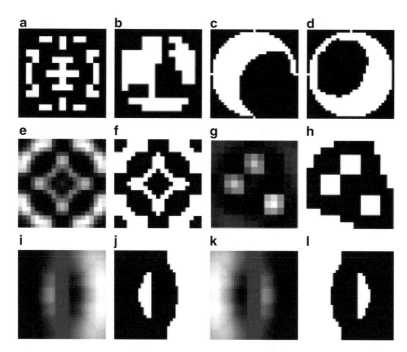

Fig. 5.6 Examples of optimized mask patterns. (**a**) Levin's mask. (**b**) Zhou's mask, $\sigma = 0.005$. (**c**) One of Zhou's pair, Zhou 1. (**d**) One of Zhou's pair, Zhou 2. (**e**) Sellent's gray-scale mask. (**f**) Sellent's binary mask. (**g**) Sellent's asymmetric gray-scale mask. (**h**) Sellent's asymmetric binary mask. (**i**) One of Sellent's gray-scale pair. (**j**) One of Sellent's binary pair. (**k**) One of Sellent's gray-scale pair. (**l**) One of Sellent's binary pair

Zhou et al. [32], which we call Zhou's mask pair, is particularly important. Zhou's mask pair, shown in Fig. 5.6c, d, is designed so that the masks complement each other in terms of spectral content and thus the information is preserved. In other words, the contradictory requirements set by depth estimation and image restoration are satisfied simultaneously. Moreover, both masks have asymmetric pattern, and this solves the sign problem mentioned in Sect. 5.4.1.

5.4.1.2 Analytic Search

The analytic search enables mask optimization in gray-scale, by expressing the evaluation criteria as an analytic function. Let us explain expression of the evaluation criterion as an analytic function, by an example of optimizing a single mask for depth estimation purpose [23]. As mentioned in Sect. 5.3.2, PSFs for different depths define different subspaces. For the optimal mask, the corresponding PSFs for different depths should be easily discriminable, so are the subspaces defined by PSFs. This discrimination capability can be measured by projecting an image onto two different subspaces and taking the difference of residues. In order to apply this idea for evaluating mask patterns, a sufficiently large set of natural sharp images F_{train}^0 is needed, to eliminate the influence of a specific image being projected. The distance between two subspaces, corresponding to PSFs $\boldsymbol{h}^{c,d_i}$ and $\boldsymbol{h}^{c,d_j}$, can then be defined as $\sum_{f_M^0 \in F_{\text{train}}^0} \left\| \mathscr{P}_i \boldsymbol{g}_M^{c,d_i} - \mathscr{P}_j \boldsymbol{g}_M^{c,d_j} \right\|_2^2$. The depth discrimination capability of each mask can be expressed as a summation of distances between all pairs of corresponding PSFs derived for a finite set of depths $\mathscr{K}$. Thus,

$$E(c) = \sum_{f_M^0 \in F_{\text{train}}^0} \sum_{d_i \neq d_j} \left\| \mathscr{P}_i \boldsymbol{g}_M^{c,d_i} - \mathscr{P}_j \boldsymbol{g}_M^{c,d_j} \right\|_2^2, \forall d_i, d_j \in \mathscr{K}. \tag{5.35}$$

Recalling the aperture superposition principle mentioned in Sect. 5.2.1, an image blurred by a PSF corresponding to depth d can be represented as

$$\begin{aligned} \mathscr{P} \boldsymbol{g}_M^{\alpha,d} &= \mathscr{P} \left[\boldsymbol{g}_M^{1,d}, \ldots, \boldsymbol{g}_M^{n^2,d} \right] \boldsymbol{\alpha} \\ &= \mathscr{P} \boldsymbol{N}^d \boldsymbol{\alpha}, \end{aligned} \tag{5.36}$$

where each column of $\boldsymbol{N}^d$ is an image vector (obtained by ordering the elements of the image into a one-dimensional array) corresponding to an elementary aperture. Then, the function given by Eq. (5.35) can be simplified as

$$E(\boldsymbol{\alpha}) = \boldsymbol{\alpha}^T \boldsymbol{M}_{\text{CA}} \boldsymbol{\alpha}, \tag{5.37}$$

where $\boldsymbol{M}_{\text{CA}} = \left\| \mathscr{P}_i \boldsymbol{N}^{d_i} - \mathscr{P}_j \boldsymbol{N}^{d_j} \right\|_2^2$. Taking into account the optical efficiency, the mask pattern design can be written as a constraint optimization problem,

$$\boldsymbol{\alpha}^* = \arg\max_{\boldsymbol{\alpha}} (\boldsymbol{\alpha}^T \boldsymbol{M}_{\text{CA}} \boldsymbol{\alpha} + \lambda \|\boldsymbol{\alpha}\|_2^2), \tag{5.38}$$

where $\|\boldsymbol{\alpha}\|_1 = 1$ and $\alpha_i \geq 0$. An example single mask pattern which is of resolution 21×21 and optimized for depth estimation purpose is shown in Fig. 5.6e [23]. Similar procedures have also been used in the case of a pair mask optimization, for depth estimation purpose. An example pair of resolution 33×33 is shown in Fig. 5.6i, k [23], where, in this case, both masks are restricted to be binary and thus they are found via binary optimization. For some practical reasons, e.g., manufacturing, the optimized gray-scale single mask patterns or mask pattern pairs can also be reduced to binary mask patterns by setting a threshold. Such examples are shown in Fig. 5.6f, h, j, l.

5.5   Combination of Coded Aperture and Stereo

The stereopsis describes the human capability of sensing 3D information based on two perspective views. As the stereopsis cue (disparity cue) is a primary cue in human vision, it is also the most popular depth cue in computer vision which is utilized by a pair of cameras. In this section, we analyze the relation between the defocus blur cue and the disparity cue and then, more importantly, investigate their joint utilization in stereo cameras which are equipped with coded apertures (coded aperture stereo cameras) to explore possible improvements in depth estimation.

5.5.1   Disparity Cue

Since the stereo vision has been extensively studied in computer vision community, here we include only the relevant information for this section, for more information, please refer to [25]. As a binocular cue, the disparity cue is encoded in two views. The binocular disparity refers to the relative location difference in images of an object captured by left and the right eyes, or cameras. In a stereo camera setup, two cameras are usually placed parallel to each other, separated by a distance B which is called the baseline, as shown in Fig. 5.7. The stereo matching algorithms find correspondences between stereo images. Having matched the scene points in two camera views, one can then obtain the depth information via triangulation as illustrated in Fig. 5.7. The relation between depth d and disparity $disp$ is given as

$$\text{disp} = \frac{fB}{d}, \tag{5.39}$$

where f is the distance between the center of projection and image plane in the pinhole camera model. This relation reveals that under the stereo camera setup, the disparity is inversely proportional to the depth. More importantly, if the same discrimination criterion is applied to the whole depth range, the depth resolution provided by the disparity cue decreases as the depth increases. As a consequence of that, the disparity cue has a working range in computer vision, which is also the case in human visual system.

5.5.2   Relation and Interaction Between the Disparity and Defocus Blur Cues

Based on the analyzes done in [22], the defocus blur cue and disparity cue share the same principle but they differ in scales. The scale of the monocular defocus blur cue is determined by the lens aperture diameter, which corresponds to baseline in

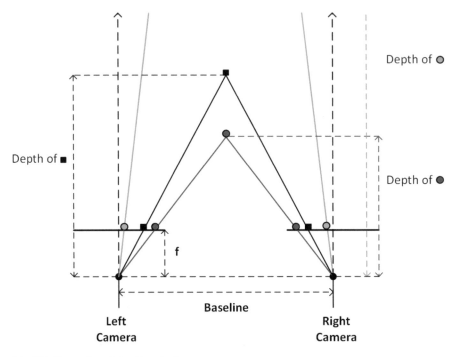

Fig. 5.7 Illustration of the disparity in stereo cameras

depth-disparity relation, as can be observed by comparing Eqs. (5.39) and (5.24). Thus, considering that the baseline is usually larger than the lens aperture diameter in most of the computer vision applications, it can be stated that stereo applications usually produce more accurate depth information than that ones utilizing the defocus blur cue. It has also been experimentally shown that the two depth cues perform in the same way, besides the scale [26, 27]: when two cameras are almost focused at the infinity, the relation between the disparity and the degree of defocus blur is linear, and the slope can be inferred as the ratio between lens aperture diameter and baseline.

Although the defocus blur and disparity cues can be treated as geometrically equivalent, besides the scale, the idea of combining defocus blur and disparity cues for improved depth information is still an interesting topic and it has actually found considerable attention in the literature [8, 20, 21, 27, 28]. The motivation behind those studies is mainly that defocus blur and disparity cues can provide complementary information.

In the following sections, we investigate possible improvements in depth estimation that can be achieved by using stereo cameras with masks, based on the work presented in [30]. Section 5.5.3 includes analysis of several integrated systems where both the defocus blur and disparity cues are available so that possible complementary information can be jointly utilized. Section 5.5.4 addresses the practical difficulty of changing the mask for coded aperture approaches employing different masks. More specifically, a single shot multiple coded aperture system is discussed.

5.5.3 Integrated Systems

In this part, several stereo camera systems with masks are analyzed as integrated systems where both the defocus blur cue and the disparity cue are available so that coded aperture and stereo vision-based depth estimation methods can work simultaneously and independently. Two different cases are considered. In the first case, both cameras are equipped with the same mask, e.g., Levin's mask shown in Fig. 5.6a. In the second case, different masks are inserted to two cameras, e.g., Zhou's mask pair shown in Fig. 5.6c, d. The motivation of having such integrated systems is examined based on the following questions. How does equipping the stereo camera with masks affect the performance of the conventional stereo matching? Can coded aperture provide useful information in situations where the stereo matching fails? Is it worth introducing masks into the stereo camera system in such cases? In order to answer those questions, a 3D scene consisting of three fronto-parallel planes, two of which are connected with a slanted plane, is considered, as shown in Fig. 5.8a. Repetitive patterns and strips are used as problematic textures, whereas gravel and rabbit's fur are employed as good textures for stereo matching. The two cameras of stereo system are assumed to be identical having a lens of 35 mm focal length. The baseline is set to be 5 cm. Both cameras are focused at 1.5 m. The captured monochromatic images are simulated by superposing the PSFs, which are found by using Eq. (5.20), corresponding to scene points, where the wavelength is taken as 534 nm. An example image from the left view in the problematic texture case and an image from the right view in the good texture case are shown in Fig. 5.8b, c, respectively.

The effect of inserting masks in the performance of stereo matching is tested by applying the same stereo matching algorithm [1] to stereo image pairs captured by stereo cameras with different sets of mask pairs (as noted in Fig. 5.9) in both the problematic texture and the good texture cases. The resulting depth maps are compared with the ground truth depth map, and the accuracies are shown in Fig. 5.9. One can infer from the results that when two identical masks are used, the influence on the performance of stereo matching is not severe [30]. It is worth to note that the employed stereo matching algorithm is a simple one. Therefore, the destructive influence of inserting masks could actually be reduced by using a more advanced stereo matching algorithm which is more robust to defocus blur.

Having cameras equipped with masks, one can now use coded aperture-based depth estimation by utilizing the defocus blur cue. For the problematic texture case, the results obtained by employing the restoration free algorithm [16] and

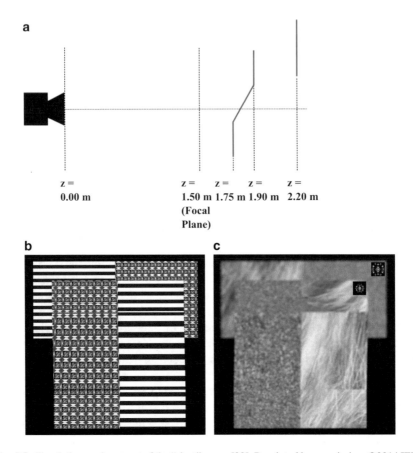

Fig. 5.8 Simulation environment of the "slant" scene [30]. Reprinted by permission. ©2014 IEEE. (**a**) The arrangement of the "slant" scene. (**b**) A *left view image* captured with pinhole aperture for the problematic texture case. (**c**) A *right view image* captured with Levin's mask for the good texture case, and two example PSFs (scaled by a factor of 3 for visualization) at depth $d = 1.9\,\text{m}$ and $d = 2.2\,\text{m}$ are shown as well

restoration-based algorithm [32], for Levin's mask and Zhou's mask pair from a single view, respectively, together with the result obtained by stereo matching in pinhole aperture case, are given in Fig. 5.10. It can be observed that it is possible to get more reliable depth information by DfD approaches than the stereo matching, which is valuable, since this information can be utilized to complement the stereo matching result. Similar results have been reported by Takeda et al. [27].

Although the occlusions and specular scene issues are not analyzed here, it is worth to also point out those cases where the depth information that is provided by coded aperture-based depth estimation (from one single view of such integrated systems) could be valuable.

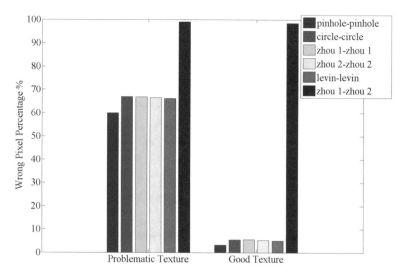

Fig. 5.9 The error percentage of stereo matching for different aperture masks, for both the problematic texture case and the good texture case [30]. Reprinted by permission. ©2014 IEEE

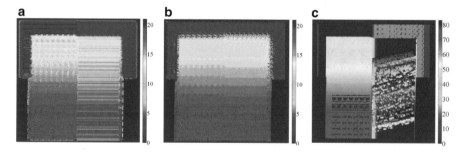

Fig. 5.10 Results produced by three algorithms for the problematic texture case [30]. Reprinted by permission. ©2014 IEEE. (**a**) The depth map in PSF scales produced by the algorithm in [16]. (**b**) The depth map in PSF scales produced by the algorithm in [32]. (**c**) The depth map in disparity values produced by stereo matching

Based on the above observations, two integrated systems have been proposed in [30], as illustrated in Fig. 5.11a, b. In the first system, two cameras are both equipped with Levin's mask, while in the second system there is a pair of images captured on both views with Zhou's mask pair. In both systems, coded aperture and stereo matching can both work independently with minimal influences to each other. When these two information sources produce complementary results, they can be merged, e.g., by using Markov random fields [28], to improve the quality of the depth map.

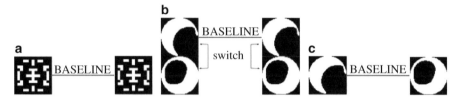

Fig. 5.11 Three proposed camera systems [30]. Reprinted by permission. ©2014 IEEE. (**a**) Stereo cameras with Levin's mask. (**b**) Two stereo cameras with Zhou's mask pair. (**c**) Stereo cameras with Zhou's mask pair

5.5.4 A Single Shot Multiple Coded Aperture System

Systems utilizing multiple coded apertures are of interest for some strong reasons. First, when only a single image is available for a camera with a particular mask, usually it is hard (or impossible) to restore the scene information of spectral content around the zero-crossings (or regions of very low amplitude) of the spectrum of the corresponding PSF. This is especially problematic for restoration-based approaches discussed in Sect. 5.3.1. However, if multiple images are captured with different masks such that those mentioned spectral regions are non-overlapping, the restoration can be successfully applied with the available complementary information. The success of restoration is particularly high for those images captured with complementary masks, such as Zhou's mask pair. Second, as mentioned in Sect. 5.4.1.1, a single mask can hardly have desired properties for both depth estimation and image restoration simultaneously, since they are contradictory. With properly designed complementary masks, the desired properties can be, however, satisfied simultaneously for both problems. Those mentioned reasons explain the benefits of using multiple coded apertures, and thus form a solid motivation to develop and use multiple masks systems.

In case multiple masks are used, it is required that corresponding images are captured from the same view to guarantee that images are well aligned. Several methods have been reported to satisfy that requirement. One way is to switch the lens mounts (with different masks) manually during capturing, and then correct the misalignment introduced during switching by using affine transformation [32]. A liquid crystal array (LCA) [13] or a liquid crystal on silicon (LCoS) [18] can be employed to make a programmable aperture camera, which then eliminates the necessity of switching the lenses. A beam splitter can also be employed to create two identical views for different masks.

There exist some practical difficulties in application of approaches mentioned above, such as the necessity of user intervention or requirement of complicated modifications in the camera system. A multi-view coded aperture system has been proposed in [30] as an alternative approach. An example arrangement is shown in Fig. 5.11c for Zhou's mask pair. Compared to the aforementioned other methods, this system has minimal modification on the lens system and does not require user intervention. The requirement of aligned images is, however, violated, since now two images are captured from different views. Therefore, a post-processing is necessary. If two images are shifted by the correct disparity value, pixels from corresponding depth can be aligned. Then, the requirement is satisfied for this particular depth and thus DfD algorithms can be applied. In order to align all pixels the same operation is to be repeated for all possible depths. In most practical cases, DfD-based depth estimation methods search the correct depth in a set of pre-sampled discrete depth values, where the search space of depths can be, naturally, chosen such that two consecutive corresponding PSFs differ in scale by one sensor pixel. Similarly, the search space for disparity values is discrete and finite, and mostly two consecutive depths correspond to two different disparity values which differ by one pixel. Hence, recalling the scale difference between defocus blur and disparity cues mentioned in Sect. 5.5.2, we end up with a multi-to-one relation between the disparity values and PSF scales for this scenario. Thus, the search space now includes disparity-PSF pairs, where for each single candidate PSF there exist multiple pairs including different disparity values. A modified version of Eq. (5.29), which is referred to as stereo version of Zhou's algorithm, can be utilized to find the correct disparity-PSF pair as [30]

$$D_M^*[l], \mathrm{Disp}_M^*[l], f_M^*[l] = \arg\min_{d_k, \mathrm{disp} \hat{f}_M^k} \sum_L \sum_{n=1}^2 \left(|g_{M_n}^{\mathrm{disp}} - h^{c_n, d_k} \otimes \hat{f}_M^k|_2^2 \right), \qquad (5.40)$$

where a pair of images, which are taken from two different views at the same time, are employed and the disparity range can be traced, e.g., by keeping the first image as it is and shifting the second image by the number of pixels in the search space. Similar algorithm has also been used in [27], where the masks utilized in stereo cameras have been same though.

The stereo coded aperture system utilizing Zhou's mask pair and stereo version of Zhou's algorithm are tested with the same simulated scene illustrated in Fig. 5.8a, for the case of problematic texture. Comparing the resulting depth maps shown in disparity and PSF scales in Fig. 5.12a, b, respectively, with Fig. 5.10b, one can observe that the stereo version of Zhou's algorithm produces as good depth map as the original Zhou's algorithm. From stereo point of view, on the other hand, an important observation is that the coded aperture stereo system produces reliable information in the problematic case where stereo matching fails for conventional stereo images. Nevertheless, the inherent occlusion problem of stereo vision also presents for the stereo coded aperture system. It should, thus, be noted as a drawback of using multiple coded aperture in stereo camera arrangement.

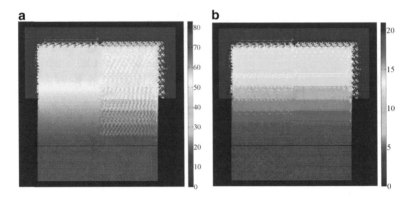

Fig. 5.12 The results produced by the stereo version of Zhou's algorithm for the "slant" scene with the problematic texture [30]. Reprinted by permission. ©2014 IEEE. (**a**) Depth map in disparity values. (**b**) Depth map in PSF scales

5.6 Conclusion

DfD approach has been shown to be an attractive alternative for acquiring scene depth information, especially, in cases where only a single photographic camera is available. Its passive nature makes it favorable in power-constrained devices. Coded aperture concept has been considered as a way of improving the performance of DfD methods by simply inserting a coded mask in the camera aperture position. Such coded aperture cameras not only improve the DfD performance but they also enable sensing the 3D scene depth information from only a single captured image, which might make them useful for dynamic scenes.

Combination of coded aperture and stereo vision by employing coded aperture stereo cameras is another case where DfD approach has been shown to be useful. It has been demonstrated that having such cameras, the stereo vision-based depth estimation can be complemented with the valuable information obtained by DfD. Defocus blur cue has been particularly shown to be useful in the cases where stereo matching suffers from the correspondence problem, e.g., repetitive textures.

There are also some deficiencies noted about the DfD approach and coded aperture cameras used for DfD purpose. The depth accuracy that can be provided by DfD methods has been shown to be inherently limited by the effective width of the camera aperture. Similarly, in stereo vision the depth accuracy is mainly determined by the extent of the baseline. In most applications, however, the aperture sizes of the cameras employed in DfD are significantly smaller than the typical baselines in stereo camera setups, for obvious practical reasons. Therefore, the depth accuracy that can be provided by DfD is usually worse compared to stereo vision. Regarding the coded aperture cameras employed for DfD purpose, on the other hand, it should be pointed out that the codes utilized in those cameras are usually optimized for a specific depth range and certain camera parameters. Thus, their performance is questionable for other scenarios. Another point that should be noted about the optimality of the designed mask is that most of those masks have been optimized within a search space of coarse resolution, and mostly binary transmittance, codes. The intention has been to get codes within a tractable search space and avoid diffraction effects. In this respect, the designed codes are actually suboptimal. Optimal mask design is, therefore, still an interesting and open research topic.

References

1. Abbeloos W (2010) Real-time stereo vision. Master's thesis, Karel de Grote-Hogeschool University College
2. Aggarwal M, Ahuja N (2002) A pupil-centric model of image formation. Int J Comput Vis 48(3):195–214
3. Bertero M, Boccacci P (1998) Introduction to inverse problems in imaging. CRC Press, Boca Raton
4. Burge J, Geisler WS (2011) Optimal defocus estimation in individual natural images. Proc Natl Acad Sci 108(40):16849–16854
5. Dowski ER, Cathey WT (1994) Single-lens single-image incoherent passive-ranging systems. Appl Opt 33(29):6762–6773
6. Dowski ER, Cathey WT (1995) Extended depth of field through wave-front coding. Appl Opt 34(11):1859–1866
7. Favaro P, Soatto S (2005) A geometric approach to shape from defocus. IEEE Trans Pattern Anal Mach Intell 27:406–417
8. Gheta I, Frese C, Heizmann M, Beyerer J (2007) A new approach for estimating depth by fusing stereo and defocus information. In: GI Jahrestagung (1)'07, pp 26–31
9. Goodman J (2004) Introduction to Fourier optics, 3rd edn. Roberts and Company, Englewood
10. Hiura S, Matsuyama T (1998) Depth measurement by the multi-focus camera. In: Proceedings of 1998 IEEE computer society conference on computer vision and pattern recognition, pp 953–959
11. Lanman D, Raskar R, Taubin G (2008) Modeling and synthesis of aperture effects in cameras. In: Proceedings of the fourth eurographics conference on computational aesthetics in graphics, visualization and imaging, eurographics association. Computational Aesthetics'08, Aire-la-Ville, pp 81–88
12. Levin A, Fergus R, Durand F, Freeman WT (2007) Image and depth from a conventional camera with a coded aperture. In: ACM SIGGRAPH 2007 papers, SIGGRAPH '07. ACM, New York
13. Liang CK, Lin TH, Wong BY, Liu C, Chen HH (2008) Programmable aperture photography: multiplexed light field acquisition. In: ACM SIGGRAPH 2008 papers, SIGGRAPH '08. ACM, New York, pp 55:1–55:10
14. Lin J, Ji X, Xu W, Dai Q (2013) Absolute depth estimation from a single defocused image. IEEE Trans Image Process 22(11):4545–4550
15. Liu C, Freeman W, Szeliski R, Kang SB (2006) Noise estimation from a single image. In: 2006 IEEE computer society conference on computer vision and pattern recognition, vol 1, pp 901–908. doi:10.1109/CVPR.2006.207
16. Martinello M, Favaro P (2011) Single image blind deconvolution with higher-order texture statistics. In: Cremers D, Magnor M, Oswald M, Zelnik-Manor L (eds) Video processing and computational video. Lecture notes in computer science, vol 7082. Springer, Berlin/Heidelberg, pp 124–151
17. Masia B, Presa L, Corrales A, Gutierrez D (2012) Perceptually optimized coded apertures for defocus deblurring. Comput Graph Forum 31(6):1867–1879

18. Nagahara H, Zhou C, Watanabe T, Ishiguro H, Nayar S (2010) Programmable aperture camera using lcos. In: Daniilidis K, Maragos P, Paragios N (eds) Computer vision - ECCV 2010. Lecture notes in computer science, vol 6316. Springer, Berlin/Heidelberg, pp 337–350
19. Pentland AP (1987) A new sense for depth of field. IEEE Trans Pattern Anal Mach Intell 9(4):523–531
20. Rajagopalan A, Chaudhuri S, Mudenagudi U (2004) Depth estimation and image restoration using defocused stereo pairs. IEEE Trans Pattern Anal Mach Intell 26(11):1521–1525
21. Saxena A, Schulte J, Ng AY (2007) Depth estimation using monocular and stereo cues. In: IJCAI, vol 7
22. Schechner Y, Kiryati N (2000) Depth from defocus vs. stereo: How different really are they? Int J Comput Vis 39(2):141–162
23. Sellent A, Favaro P (2014) Optimized aperture shapes for depth estimation. Pattern Recognit Lett 40:96–103
24. Sellent A, Favaro P (2014) Which side of the focal plane are you on? In: 2014 IEEE international conference on computational photography (ICCP), pp 1–8
25. Snowden R, Thompson P, Troscianko T (2012) Basic vision: an introduction to visual perception, revised edn. Oxford University Press, Oxford, 424 p.
26. Takeda Y, Hiura S, Sato K (2012) Coded aperture stereo: for extension of depth of field and refocusing. In: VISAPP 2012 - Proceedings of the international conference on computer vision theory and applications, vol 1, pp 103–111
27. Takeda Y, Hiura S, Sato K (2013) Fusing depth from defocus and stereo with coded apertures. In: 2013 IEEE conference on computer vision and pattern recognition (CVPR), pp 209–216
28. Tao M, Hadap S, Malik J, Ramamoorthi R (2013) Depth from combining defocus and correspondence using light-field cameras. In: 2013 IEEE international conference on computer vision (ICCV), pp 673–680
29. Wang C (2015) Design and analysis of coded aperture for 3d scene sensing. Master's thesis, Tampere University of Technology
30. Wang C, Sahin E, Suominen OJ, Gotchev AP (2014) Depth estimation by combining stereo matching and coded aperture. In: IEEE conference on visual communications and image processing (VCIP), pp 291–294
31. Zhou C, Nayar S (2009) What are good apertures for defocus deblurring? In: 2009 IEEE international conference on computational photography (ICCP), pp 1–8
32. Zhou C, Lin S, Nayar S (2009) Coded aperture pairs for depth from defocus. In: 2009 IEEE 12th international conference on computer vision, pp 325–332
33. Zhou C, Lin S, Nayar S (2011) Coded aperture pairs for depth from defocus and defocus deblurring. Int J Comput Vis 93(1):53–72
34. Zhu X, Cohen S, Schiller S, Milanfar P (2013) Estimating spatially varying defocus blur from a single image. IEEE Trans Image Process 22(12):4879–4891

Chapter 6
Depth Map Coding for 3DTV Applications

Carl James Debono, Sérgio Faria, Luís Lucas, and Nuno Rodrigues

Abstract The communication of multi-view videos requires the transmission and storage of huge amounts of data. To reduce this storage and bandwidth requirements, a reduced set of videos together with the depth information can be used. The 3D geometry information in the depth maps is used in conjunction with the texture information to generate any intermediate view between two received video streams. Unlike texture information, depth data is characterized by large homogeneous areas and sharp edges, where the latter define object boundaries. These different features imply that encoding of depth maps with the standard texture encoder might not be optimal and thus different methods can be used to improve the coding efficiency of depth maps. This chapter presents coding algorithms used for the compression of depth maps in 3DTV applications.

6.1 Introduction

Three-dimensional (3D) television (TV) applications, such as multi-view and free-view systems, demand huge amounts of data transmission. The bandwidth needed for transmission is prohibitive in most of the channels available today. A possible 3D video format that can allow drastic reduction of the data needed for transmission is the multi-view video-plus-depth. The depth map video provides geometry information about the scene, which is used by the view synthesis algorithms to generate virtual views in between the transmitted camera views. The process used is based on depth-image-based rendering (DIBR [1]) which projects the reference views onto the required virtual view.

The depth maps are not displayed at the receiver but the synthesized views generated depend on the quality of these depth maps. Furthermore, the depth maps

C.J. Debono (✉)
Department of Communications and Computer Engineering, University of Malta, Msida, Malta
e-mail: c.debono@ieee.org

S. Faria • L. Lucas • N. Rodrigues
Instituto de Telecomunicações, Leiria, Portugal

Universidade Federal do Rio de Janeiro, Rio de Janeiro, Brazil
e-mail: sergio.faria@co.it.pt; luis.lucas@smt.ufrj.br; nuno.rodrigues@co.it.pt

© Springer Science+Business Media New York 2017
A. Kondoz, T. Dagiuklas (eds.), *Connected Media in the Future Internet Era*,
DOI 10.1007/978-1-4939-4026-4_6

are composed of homogeneous areas and sharp edges, where the edges define the difference in depth and thus their preservation is very important. This suggests that coding of the depth maps must be performed diligently. This chapter explores the state-of-the-art coding schemes employed in depth map coding and provides insight into future directions in this field.

6.2 Efficient Depth Map Coding in 3D Video

The high-efficiency video coding (HEVC) [2, 3] is the most recent standard for video compression, providing significant performance gains over its predecessor H.264/AVC standard [4], with an average of 50 % bitrate reduction for the same perceptual video quality. As most video coding standards, HEVC uses a block-based hybrid coding approach, based on intra- and inter-frame prediction methods for efficient coding of spatial and temporal similarities, and 2D transform-based residue coding. Despite its advantage for generic videos, HEVC presents some issues when used for depth map coding.

In this section we present an overview of the most relevant algorithms for depth map coding proposed in the literature. Depth-image-based rendering (DIBR) is firstly introduced, since it is a common process in many multi-view video coding algorithms. The depth map coding schemes described in the later subsections include the upcoming 3D extension of the state-of-the-art HEVC standard, known as 3D-HEVC [5–7], which adds support for efficient depth map coding. In the case of 3D-HEVC, depth maps are jointly encoded with texture video (also referred as texture or color video), which allows exploiting an additional source of redundancy, by encoding the similarities between both components. Other presented algorithms include alternative coding techniques to the transform-based coding, in order to better preserve the sharp edges that characterize depth maps.

6.2.1 Depth-Image-Based Rendering

The rendering technique used affects the quality of the virtual views. One of the most popular techniques is DIBR and it is the technique employed by the Motion Picture Experts Group (MPEG) as the reference synthesis framework for free-viewpoint video architectures, which relies on the multi-view video-plus-depth format. The view synthesis reference software (VSRS) released by the ad hoc group on 3D audio and visual (3DAV) of MPEG is based on DIBR [1].

A block diagram of the DIBR algorithm is shown in Fig. 6.1. It requires two reference views and their associated depth maps to generate any virtual view between these two references. The two nearest reference views to the required virtual viewpoint are selected from the multi-view sequence and warped [8].

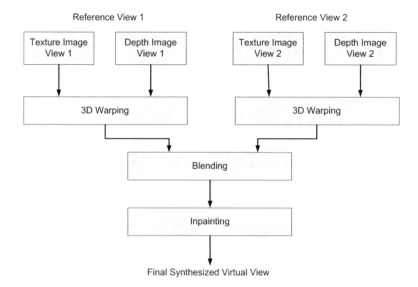

Fig. 6.1 Block diagram of the DIBR algorithm

The warped images generated from the two views are then blended to form the new virtual position [9]. Since some disoccluded regions and holes will still remain, inpainting is applied to fill the missing data [8].

The warping process is done pixel by pixel and consists of two parts, where during the first part the pixels in the reference view are back-projected to the 3D world coordinate space using the depth values of the reference view and the camera parameters. For a pixel (x_r, y_r) this is given by [1]

$$
\begin{pmatrix} X_W \\ Y_W \\ Z_W \end{pmatrix} = R_r^{-1} \left(z_r K_r^{-1} \begin{pmatrix} x_r \\ y_r \\ 1 \end{pmatrix} - \mathbf{t}_r \right)
$$

where (X_W, Y_W, Z_W) are the world coordinates, K_r is a 3×3 intrinsic matrix, R_r is a 3×3 rotation matrix, z_r is the depth value, and $\mathbf{t}_r$ is a 3×1 translation vector of the reference camera. In the second part of the process, these 3D points are forward-projected to the desired image view using the virtual camera parameters [10]. The resulting pixel position in the virtual view (x_v, y_v) is found using [1]

$$
z_v \begin{pmatrix} x_v \\ y_v \\ 1 \end{pmatrix} = K_v \left(R_v \begin{pmatrix} X_W \\ Y_W \\ Z_W \end{pmatrix} + \mathbf{t}_v \right)
$$

where K_v is a 3×3 intrinsic matrix, R_v is a 3×3 rotation matrix, z_v is the depth value, and $\mathbf{t}_v$ is a 3×1 translation vector of the virtual camera.

The virtual camera parameters are unknown and thus must be estimated from the physical cameras. The intrinsic matrix and the translation vector, c, can be easily found using linear interpolation [10] using [11]

$$K_v(\alpha) = (1 - \alpha)K_1 + \alpha K_2$$

$$c_v(\alpha) = (1 - \alpha)c_1 + \alpha c_2$$

where subscripts 1 and 2 refer to view 1 and view 2 in Fig. 6.1, and α is the position-dependent parameter. The vector c is related to t through

$$t = -Rc$$

Linear interpolation cannot be used for the rotation matrix as orthonormality will be lost. Two common methods used to compute the virtual camera's rotation matrix are Euler angles and spherical linear interpolation (SLERP). The interpolation using SLERP of q that lies between q_0 and q_1 is given by [12]

$$q = \text{SLERP}(q_0, q_1, \alpha) = \frac{\sin[(1 - \alpha)\Omega]}{\sin \Omega}q_0 + \frac{\sin(\alpha\Omega)}{\sin \Omega}q_1$$

where q_0 and q_1 are two unit quaternions that represent the rotation matrices of the reference cameras, q is the quaternion of the rotation matrix of the virtual camera, and Ω is the angle between q_0 and q_1. The conversion between quaternion and the rotation matrix is done through [12]

$$R = \begin{pmatrix} 1 - 2y^2 - 2z^2 & 2xy + 2wz & 2xz - 2wy \\ 2xy - 2wz & 1 - 2x^2 - 2z^2 & 2yz + 2wx \\ 2xz + 2wy & 2yz - 2wx & 1 - 2x^2 - 2y^2 \end{pmatrix}$$

where $q = [w, (x, y, z)]$, w is the scalar part of the quaternion, and (x, y, z) is the imaginary part.

Most DIBR algorithms are based on inverse warping since it can provide better rendering quality compared to forward warping [1]. The two reference depth maps are warped to the virtual view and a median filter is applied to fill the holes generated by this process. During the texture mapping process, the processed depth images are inversely warped to map the original texture images to the virtual view. Disocclusion regions are filled from the respective views and the boundary noise is removed. This cleans the ghosting artifacts in the generated views by detecting the background boundary, determining the boundary noise region, and replacing the texture of this region from the other warped image [13]. The texture images are then blended and any remaining holes are inpainted.

In [14], the reference depth maps are pre-processed to reduce errors introduced by stereo matching algorithms. The processing includes temporal filtering, compensation for errors, and spatial filtering. Inverse warping is also used in [14]. Illumination and color mismatches between the warped images are corrected using a histogram matching algorithm, which reduces color inconsistencies in the synthesized view. The two images are then blended and the holes filled through inpainting.

An illumination compensation technique is applied in [15] to reduce color discontinuities. This is done by adjusting the brightness of the auxiliary view to match the main view before the blending process. This results in better visual quality of the synthesized views.

The warped depth maps are processed by median and bilateral filters before inverse warping in [16]. The median filter fills the small holes that result because of the integer approximations of warped coordinates while the bilateral filter smoothens the images while preserving the edges. The disoccluded regions in the virtual texture view are dilated before blending and inpainting.

Depth map pixels at edges are detected and are not warped in [17], thus reducing unreliable data from the warping operations. The reference depth map is converted to a binary image and dilated. Subtracting these binary images from this results in detection of the unreliable boundary, as shown in Fig. 6.2. The brightness adjustment algorithm uses the depth data to exclude foreground objects from biasing the results. Furthermore, an adaptive blending technique chooses between alpha-blending and base-plus-assistant blending, depending on the current virtual viewpoint. If this is closer to a reference view base-plus-assistant blending is used, while if it is further away alpha-blending is employed [17]. Since the object's boundary pixels are not warped, an average filter is applied to these edges after inpainting to smoothen them and provide a more natural look.

Other DIBR techniques include the refinement of virtual views through pixel classification, graph cuts, and depth-based inpainting [18]. The perceived depth quality and visual comfort in stereoscopic images are improved through the application of stereoacuity before rendering in [19]. Furthermore, a just noticeable depth difference (JNDD) model and saliency analysis are used in [20] to enhance the perception of the rendered views.

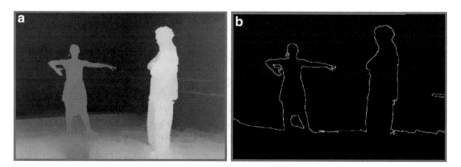

Fig. 6.2 Detection of the unreliable edges, (**a**) the depth map from the *Ballet* sequence and (**b**) the detected unreliable boundary of (**a**) [17]

6.2.2   Treatment of Depth Data in MV-HEVC and 3D-HEVC

In general, depth maps are characterized by large homogeneous regions, separated by sharp edges at object boundaries, thus differing from typical texture video contents. This means that more particular coding tools may be investigated for depth map signals. In 3D video-plus-depth format, depth map views represent the same scene as the video, because the video acquisition is performed from the same viewpoints. Therefore, a strong similarity may exist between these components, which can be exploited by the coding algorithms. These facts indicate that adequate algorithms for the compression of depth maps may be developed. Nevertheless, straightforward solutions that provide independent compression of depth maps may also be used, based on the existing video coding technologies.

6.2.2.1   Depth Map Coding Using MV-HEVC

A trivial approach for depth map coding would be the use of the state-of-the-art multi-view extension of the high-efficiency video coding standard (MV-HEVC) [5–7]. The MV-HEVC has been developed for efficient compression of texture-based multi-view video (MVV) data, by exploiting the similarities existing between multiple video views, captured from the same 3D scene at slightly different spatial positions. These inter-view dependencies are mainly represented by disparity-compensated prediction (DCP), as in the case of the multi-view video coding (MVC) extension of the previous H.264/MPEG-4 AVC standard.

DCP is implemented by modifying the high-level syntax of the HEVC standard. In practice, the reference picture list of the current view, which originally contains the temporal frames for motion-compensated prediction, is augmented with the inclusion of already encoded frames from other views at the same time instances. In this way, the block-matching-based search procedure may adaptively find the optimal predictor, either from temporal or inter-view references, for each encoded block. The design of DCP provides a straightforward solution to adapt the existing HEVC technology for efficient representation of MVV data, because it does not require any changes at block-level tools.

Despite MV-HEVC provides an effective solution to exploit inter-view dependencies between depth map views, the intra-frame coding tools inherited from HEVC encoder present some issues for the compression of depth maps. These problems are mainly associated to the characteristics of depth signals, which differ from natural images. Current video coding standards use transform-based intra-coding tools that tend to generate a high number of coefficients in blocks containing sharp edges. The main problem is the ringing artifacts generated around those edges caused by the quantization of the transform coefficients. As depth maps are typically used for view synthesis purposes, these artifacts in the encoded depth maps tend to generate strong distortions in the synthesized virtual views.

6.2.2.2   Depth Map Coding Using 3D-HEVC

The 3D-HEVC algorithm [5–7], whose standardization was being conducted by the Joint Collaborative Team on 3D video coding extension development (JCT-3V) formed by elements from MPEG (ISO/IECJTC1/SC29/WG11 3DV ad hoc group) and ITU-T (SG 16 WP 3), is the depth-enhanced 3D video coding extension to the HEVC standard. The aim of 3D-HEVC is to efficiently encode 3D contents under video-plus-depth format, which includes both texture video and depth map views. This algorithm is built over the MV-HEVC standard that already provides an efficient compression of the texture video component, by exploiting the intra-, temporal-, and inter-view dependencies. However, for an improved rate-distortion performance, the 3D-HEVC standard introduces modifications at block-level tools, which allow exploiting correlations of motion and residue information between the multiple views. The most relevant tools include the neighboring block-based disparity vector derivation (NBDV), inter-view motion prediction, inter-view residual prediction, and illumination compensation.

The main idea of the NBDV tool is to implicitly derive the disparity vector of the block, based on its spatial and temporal neighborhood, in order to be used by other methods, namely inter-view motion prediction and inter-view residual prediction. NBDV uses spatial candidates similar to the ones used in advanced motion vector prediction (AMVP) or merge modes of HEVC plus a temporal neighboring candidate. The first disparity vector found in the spatial and temporal candidates is returned by NBDV. When the disparity vector is not included in the available candidates, a zero disparity vector is assumed. Note that this approach does not require any additional signaling symbol in the bit stream.

Regarding inter-view motion prediction, its main purpose is to include additional candidates into the list of the merge mode, which can sum up to six candidates instead of five, as defined in the single-view HEVC standard. Two possible additional candidates are the disparity vector and the reference picture index derived by the NBDV itself, or the motion vector associated to the block pointed out by the NBDV vector in the corresponding reference picture index.

The inter-view residual prediction is able to exploit similarities between the motion-compensated residue produced in two distinct views. Its procedure consists in the prediction of the motion-compensated residual signal of the current block using an already encoded view, namely the motion-compensated residual signal associated to the block that is pointed out by the NBDV disparity vector and the corresponding reference picture index.

The illumination compensation is another block-level tool proposed to improve MVV data compression. It is used to compensate luminance variations often observed between views, for instance when the cameras present calibration differences. For this reason, this tool is only applied on blocks predicted from inter-view reference pictures. The illumination compensation is achieved by using a simple linear model whose coefficients are implicitly estimated by a least-square solution based on the block neighboring reconstructed samples.

In addition to the new block-level tools for MVV coding, 3D-HEVC proposes additional intra- and inter-coding tools to better represent depth maps, and exploits the new level of dependency arising from the joint compression of texture video and depth maps, referred to as inter-component redundancy. Moreover, some tools inherited from MV-HEVC were modified or disabled, for the coding of depth maps. The modified tools include the motion and disparity compensation methods that use full-sample accuracy instead of the quarter-sample. This is because the eight-tap interpolation filters tend to create ringing artifacts in depth maps. Similarly, the estimated vectors are encoded using full-sample accuracy. The disabled tools in 3D-HEVC for depth map coding include the in-loop filters, specifically the de-blocking filter, the adaptive loop filter, and the sample-adaptive loop filter.

Regarding intra-coding methods for depth maps, 3D-HEVC maintains the directional prediction modes and transform-based residue coding present in the reference HEVC standard. However, new tools that better preserve depth map discontinuities have been introduced [21], namely the depth modeling modes (DMM), segment-wise DC coding (SDC), and single depth intra-mode. DMM takes an important role for depth map representation in 3D-HEVC standard. It consists in new intra-prediction modes that partition the depth block into two non-rectangular regions, which are approximated by constant values. The partitioning information and the mean value of both regions are the unique information required by this model. DMM may exploit some inter-component dependencies by deriving the partitioning information of the depth block from the co-located block in the corresponding texture video view. Similarly to directional modes, the residual signal calculated between the DMM model and the original depth map might be encoded using transform coding. More details about DMM can be found in Sect. 6.3.1.

The SDC mode is an alternative residual coding method to the transform coding and quantization. SDC can only be employed for prediction units (PUs) of size $2N \times 2N$. For directional intra-prediction, one segment is defined, while two segments are defined for DMM. For each segment, SDC determines the mean value of original depth samples and it uses a predicted depth value in order to form a residual value that should be transmitted using a depth lookup table (DLT). The residual DC value mapping provided by DLT allows reducing the residue bit depth, especially when original depth maps do not use the whole range of 256 depth values, e.g., quantized depth maps. The single depth intra-mode is another tool used for efficient representation of smooth areas in depth maps. It aims to approximate the depth block based on a single depth sample value that is derived from a sample candidate list. Specific sample positions in the current block neighborhood are used to derive the sample candidate list. Whenever this mode is used, no residual information is transmitted.

In relation to depth map inter-coding, several techniques are introduced in 3D-HEVC. The motion parameter inheritance (MPI) is used to predict depth maps based on the motion characteristics of the associated texture video view already encoded. The main idea behind this technique is to exploit the similar characteristics between the motion in both video and depth map signals. This is because the texture video and corresponding depth map signal capture the same scene from the same viewpoint at the same time instants. Therefore, MPI adds a new candidate into the merge list of the current depth block, which is derived from the motion information of the co-located texture block.

Inter-predicted blocks in depth maps may skip transform and quantization methods, using the inter-mode extension of the previous presented SDC method, which approximates the residual signal using a constant DC value. The inter-mode SDC is only used when the DC residue is non-null; otherwise the skip mode is a better solution. Another source of redundancy resides in the block partitioning tree of both depth and texture blocks. 3D-HEVC exploits block partition similarities by predicting the depth map quadtree from the associated texture block quadtree for inter slices.

In 3D-HEVC, the joint coding of texture video and depth maps and consequent exploitation of inter-component similarities are not uniquely used to improve the compression of depth maps, but also to compress texture video more efficiently. For this, 3D-HEVC proposes new texture video coding tools adding support for inter-component coding. In the case of texture video, the corresponding depth map view is not accessible because texture component is always encoded first. However, when there are encoded depth maps from other views, they can be exploited for the compression of the current texture view. Exploiting redundancies from different depth map views is possible, because depth maps convey disparity information that may be used for warping the depth map pictures onto the desired view position, using the camera array parameters. In this context, 3D-HEVC proposes an improved NBDV method, referred to as depth-oriented neighboring block-based disparity vector (DoNBDV), for texture video coding. This method improves the estimation of the disparity information, using depth map data, in contrast to the previous presented NBDV algorithm. Additional techniques for texture view coding in 3D-HEVC include the view synthesis prediction (VSP) and the depth-based block partitioning (DBBP).

Video coding encoders typically use a Lagrangian cost function: $J = D + \lambda R$ for mode decision. The distortion (D) and number of bits (R) associated to each candidate mode are estimated and the one that produces the smallest cost value is chosen. The λ value is used to control the video bitrate and distortion trade-off. Although the encoder procedures are not specified by video coding standards, the use of a Lagrangian function is mandatory, in order to provide the best quality for a given bitrate. In the 3D-HEVC encoder, the Lagrangian cost function is also used for depth map coding. However, as depth maps are not directly observed by the users, the coding efficiency is improved by using an alternative distortion metric that considers the distortion of the synthesized views, instead of the sum of squared differences (SSD) or sum of absolute differences (SAD) of the depth block. This method, known as view synthesis optimization, is demonstrated in 3D-HEVC encoder test model,

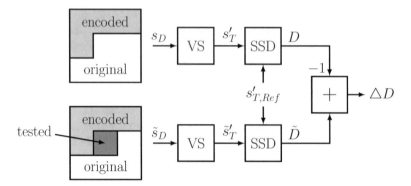

Fig. 6.3 SVDC procedure for the distorted depth block identified in bottom depth map

under two main implementations: the synthesized view distortion change (SVDC) and a model-based synthesized distortion estimation, which avoids rendering, for low-complexity distortion calculation [21].

SVDC estimates the change of distortion in the synthesized intermediate view caused by the change in the depth block due to the coding error. This process is defined as the distortion difference, $\Delta\boldsymbol{D}$, computed between two synthesized texture images, as illustrated in Fig. 6.3.

The first texture image, s'_T, is generated with a depth map that comprises the reconstructed depth values for the already encoded blocks, and uncompressed depth values for the remaining depth map blocks. The second texture image, $\tilde{s}'_T$, is synthesized from a similar depth map that only differs in the samples of the block currently being encoded, which correspond to the reconstructed samples produced by the mode under test, signaled in Fig. 6.3 by the dark pattern. The SSD for each synthesized view is estimated against a reference synthesized texture view, $s'_{T,\mathrm{Ref}}$, which is rendered from the uncompressed depth and texture video data. The whole SVDC calculation can be represented by the following expression:

$$\Delta D = \widetilde{D} - D = \sum_{(x,y)\in I}\left[\tilde{s}'_T(x,y) - s'_{T,\mathrm{Ref}}(x,y)\right]^2 - \sum_{(x,y)\in I}\left[s'_T(x,y) - s'_{T,\mathrm{Ref}}(x,y)\right]^2$$

where I represents the set of samples of the synthesized view. In order to avoid the rendering of the entire synthesized view for each tested block and save some computational complexity, 3D-HEVC uses an efficient implementation of SVDC that performs minimal rendering of the regions that are affected by the distorted depth block under test.

The use of VSO techniques in the 3D-HEVC encoder has shown to provide significant rate-distortion gains when compared to the traditional distortion metrics. Figure 6.4 compares the performance of MV-HEVC and 3D-HEVC, based on the MPEG common test conditions for 3D video core experiments, for the *shark* test sequence [22]. Two configurations of the 3D-HEVC encoder are presented in order to illustrate the effect of the VSO distortion metric.

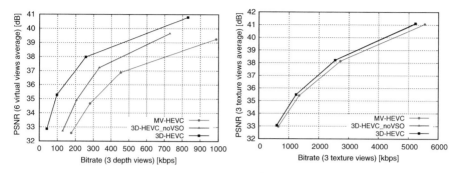

Fig. 6.4 Rate-distortion performance of *Shark* test sequence using the MV-HEVC, 3D-HEVC without VSO, and the plain 3D-HEVC encoders. *Left plot* corresponds to depth map coding results, while *right plot* shows texture video view results

As previously mentioned, since depth maps are not directly observed by the viewer, the compression efficiency of these signals is evaluated based on the distortion of the synthesized views. Thus, the performance results for depth map coding, illustrated in the left plot of Fig. 6.4, are given by the average quality of six virtual texture views, which are synthesized using three encoded depth maps and texture views. The PSNR of the intermediate texture views is plotted as a function of the total bitrate used to compress the depth map views. These results demonstrate the clear advantage of the recent depth-enhanced 3D-HEVC encoder over the MV-HEVC standard, which was not designed for efficient compression of depth maps. The results without using VSO distortion metric (3D-HEVC_noVSO curve) prove the importance of the new distortion metric for depth map coding.

In order to illustrate the effect of the joint texture and depth map coding employed in 3D-HEVC, the performance results for the texture video views are presented in the right plot of Fig. 6.4, based on the average quality of the three texture views in function of total bitrate of the same views. Comparing bitrate values between the two plots, it can also be concluded that depth maps convey much lesser bits than texture video encoded independently.

6.2.3 Depth Map Coding Schemes

As described in the previous section, the 3D-HEVC algorithm specifies several tools for efficient depth map coding. These tools have been investigated in the context of the standardization process for the compression of multi-view video-plus-depth data. However, many other coding solutions, that take into account the particular features of depth maps, have been suggested in literature. While some proposals consist of additional tools or modifications to the existing video coding standards, there are also many solutions that investigate alternative intra-frame coding paradigms, as replacement to the traditional transform-based methods.

Early versions of the coding tools used in the current 3D-HEVC algorithm can be found in literature as independent proposals for HEVC standard, such as [23–26]. The description of 3D-HEVC algorithm provided in Sect. 6.2.2 encompasses most of these techniques. In [27], a modified version of 3D-HEVC that fully replaces the transform-based residue coding and directional intra-prediction framework by an advanced geometry-based intra-prediction approach was recently proposed. This method includes plane fitting, Wedgelet modeling, inter-component prediction, and constant offset residual coding. Despite the reported coding performance not being superior to one of the 3D-HEVC encoders, it enables efficient triangular mesh extraction for scene surface representation.

Several works that propose new coding tools and modifications to the previous H.264/AVC and JPEG2000 standards for depth map coding can also be found in literature. An improved version of H.264/AVC encoder including a new edge-preserving intra-prediction mode is proposed in [28]. The purpose of this mode is to predict blocks that contain arbitrarily shaped edges. These blocks are partitioned along the edge into two disjointed regions, which are in turn approximated by constant values. The edge information is efficiently encoded using the context-adaptive binary arithmetic coding, by exploiting the structure of the previously encoded edges.

Other solutions, including new intra-prediction modes for H.264/AVC in order to better encode the edges of depth maps, have also been presented in literature, namely [29] and [30]. The work of Krishnamurthy [31] improves the JPEG2000 algorithm for depth map coding using region of interest (ROI)-based coding and reshaping the dynamic range of the depth map. The idea of ROI is to encode the most important regions for DIBR, namely edges delimiting objects in depth maps. The motivation for the reshaping of dynamic range is the fact that higher DIBR errors tend to occur in areas with lower depth.

The Platelet algorithm [32] addresses the depth coding problem in an innovative way, by proposing a complete encoder solution based on a new coding paradigm, differently from the existing video coding standards. This approach employs a quadtree decomposition over the whole input depth image, being recursively divided into blocks of different sizes. Each leaf of the quadtree is approximated using four possible modeling functions: constant, linear, Wedgelet, and Platelet functions.

Constant and linear functions are used to approximate smooth blocks, while Wedgelet and Platelet functions provide efficient representation of depth map edges. The latter two functions are piecewise functions that divide the block into two disjointed regions based on a straight line. The difference between Wedgelet and Platelet is the model which is used to approximate each region, constant, and linear functions, respectively. For each sub-block of the fully expanded quadtree, the quantized modeling coefficients and the optimal modeling function is chosen, based on a Lagrangian rate-distortion cost function. The expanded quadtree is then pruned in a bottom-up fashion, evaluating the global rate-distortion cost of each node. Experimental results show that Platelet algorithm achieves a superior performance than the H.264/AVC standard.

Alternative approaches for depth map coding are based on adaptive mesh generation. In [33], an interesting solution is proposed based on the binary triangular tree method. This method divides the input depth map into an adaptive mesh of triangles that can be represented by a binary tree. The nodes of the mesh are compared to nonuniform samples of the image. Smaller triangles are used near the depth edges in order to better approximate them, while smooth areas are approximated by larger triangles. The percentage of depth errors (PERR) inside each triangle is used to measure the reconstruction quality of the triangles, and thus determine the adaptive mesh. The uniform depth value associated to each triangle is differentially encoded using entropy coding. The binary tree is also compressed using entropy encoding. Reported results using this method have shown a lower depth error rate than JPEG2000 standard.

The use of graph Fourier transform (GFT) for the compression of depth maps has been introduced in [34], under the designation of edge-adaptive transform (EAT). These transforms tend to be more efficient in compressing depth map edges than the DCT transform. In the first proposal, GFTs are used in H.264/AVC algorithm as a residue coding method, alternative to the DCT transform. The encoder optimally chooses the transform method based on the rate-distortion cost function.

For derivation of the GFT, an edge map is first created using an edge detection algorithm over the residual block resultant from the prediction stage. Then, a graph is created based on the generated edge map. A graph connection between two neighboring samples is defined when no edge exists between them. The Laplacian of the estimated graph is computed and the GFT matrix is constructed from the eigenvectors of the graph Laplacian. The projection of the residue block signal onto the eigenvectors of the graph Laplacian provides a spectral decomposition of the residual signal in the graph. The resultant coefficients are quantized and entropy coded.

The main advantage of GFT is the ability to create fewer transform coefficients than DCT in the presence of piecewise smooth signals, such as depth maps. Particularly, it can be shown that for an image block with M constant regions, the GFT produces at most M nonzero coefficients. However, in order to reconstruct the GFT matrix in the decoder side for inverse operation, the detected edges should be explicitly encoded and transmitted as side information. This fact does not guarantee that GFT is optimal in terms of rate-distortion cost, because the transmission cost of detected depth edges may be higher than the DCT representation.

Improved versions of GFT method for the compression of piecewise smooth signals, including depth maps, have been proposed in literature. In [35], a multi-resolution GFT coding scheme is proposed for H.264/AVC standard. In this work a large number of possible GFTs are available, including unweighted and weighted graphs. Unlike [34], which directly creates graphs from detected depth edges, this approach minimizes the total representation cost, considering the cost of GFT description. Simple connectivity graphs and quantized sets of edge weights selected based on appropriate searching procedures are used in order to reduce the GFT description cost. Furthermore, low computational complexity techniques have been proposed based on the multi-resolution approach [35].

An alternative depth map coding solution based on the pattern matching paradigm has also been investigated in literature. In particular, the Multidimensional Multiscale Parser (MMP) algorithm, previously presented for generic image coding, has been applied for the compression of depth maps in [36] and [37]. Unlike transform-based coding, MMP algorithm is able to avoid filtering and ringing artifacts around edges of encoded depth maps, due to the pattern matching-based residue coding.

MMP is a block-based algorithm that performs intra-prediction and pattern-matching residue coding within a flexible block-partitioning scheme [38]. Directional prediction is used based on the H.264/AVC standard modes. The residue blocks are approximated using code words from a dictionary, adaptively created during the coding process. New dictionary code words are generated through expansions, contractions, and concatenations, among other transformations of the previous encoded residue patterns. MMP uses a highly flexible block-partitioning scheme, where the block may be recursively segmented either in horizontal or vertical direction, from 16×16 size down to 1×1. This scheme enables 25 possible block sizes, which is a far superior number than the block partitions allowed by quadtree partitioning used in most video coding standards. The optimal block-partitioning tree, intra-prediction modes, and code word indexes for residue approximation are estimated for each initial block of the depth map, based on a Lagrangian rate-distortion cost function. MMP coding performance has been shown to be higher than the one of JPEG2000, H.264/AVC, and Platelet algorithms. However, its main disadvantage is the high computational complexity associated to the pattern matching paradigm.

In the context of the MMP algorithm, new research works using the highly flexible block-partitioning scheme with linear residue approximation methods were initiated. The algorithm presented in [39] demonstrates that the flexible block-partitioning scheme combined with intra-directional prediction can be efficient for the representation of depth edges. For residue coding, a linear fitting method improved by dictionary-based reusing of past approximations was used. Unlike the MMP algorithm, the new residue coding solution uses a significantly inferior computational complexity.

In its procedure, the residue block can be approximated using a linear fitting model, selected from the dictionary or using a new fitting model, whose parameters are explicitly signaled. The optimal approximation is chosen based on a rate-distortion criterion. The advantage of the dictionary method is the reduced bitrate required to transmit one linear approximation multiple times. The dictionary is built by adding new elements corresponding to the explicitly transmitted linear models. The lower computational complexity of this solution relative to the MMP algorithm is justified by the fact that much fewer elements are added into the dictionary during the encoding process.

Figure 6.5 presents the rate-distortion results for some of the reported methods, for sequences *Ballet* (left) and *Breakdancers* (right). Since depth maps are not directly presented to the viewers, these plots illustrate the PSNR of the intermediate virtual texture image (camera 4) synthesized using the VSRS reference software

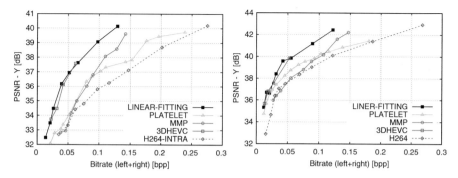

Fig. 6.5 PSNR results of the intermediate virtual image (camera 4) synthesized using the compressed depth maps and the original texture images of cameras 3 and 5, as a function of depth map's total bitrate. Results are shown for sequences *Ballet* (*left*) and *Breakdancers* (*right*)

[49] based on two compressed depth maps and original texture images of cameras 3 and 5, as a function of the bitrate used to compress both left and right depth maps. For PSNR calculations, the reference intermediate view was generated using the uncompressed depth maps and texture views. Results using depth maps encoded by the H.264/AVC standard and 3D-HEVC encoder are also presented for comparison purposes. As depth maps were independently encoded by each method, the texture-based coding tools of 3D-HEVC, such as VSO or Contour DMM, are disabled in these experiments.

Both plots show that the presented methods are able to improve the synthesis performance, when compared to the reference H.264/AVC method. However, with the exception of the linear fitting-based method of [39], all the presented methods present a lower performance than the 3D-HEVC encoder. In regard to those methods missing in Fig. 6.5, such as the GFT-based algorithms, there are no reported results presenting a competitive performance with the state-of-the-art 3D-HEVC encoder. The linear fitting-based method proposed in [39] has shown to be the only depth map coding algorithm competitive with the 3D-HEVC coding technology, being consistently superior to the Platelet and MMP algorithms.

6.3 Trends in Depth Map Coding

Encoding of depth data plays an increasingly important role in the success of free-view television (FTV) [8], as it needs to not only provide compression efficiency, but also ensure high-quality synthesized views. In this sense, the most recent algorithms include the use of depth modeling modes, coding techniques based on directional intra-prediction and flexible block partitioning, as well as post-processing techniques.

6.3.1 Depth Modeling Modes

As mentioned before, depth maps typically contain homogeneous regions and sharp edges at the boundaries of objects. This differs from the texture video content indicating that better coding tools can be used to encode this data. To support this, the 3D-HEVC standard employs additional depth modeling modes (DMM) for better representation of object edges to the available prediction modes. These modes partition the depth block into two different regions, where each region is represented by a constant partition value (CPV). Only two types of partitioning are allowed, namely Wedgelet and Contour partitioning. Wedgelet partitioning is done by dividing the depth block into two parts using a straight line. On the other hand, contour partitioning divides the depth block using arbitrary shapes. These partitioning techniques can be viewed in Fig. 6.6 [42].

Other than the value of the regions P_1 and P_2, the position of the partitioning line or the contour must be signaled to the decoder. This can be done in different ways [42], namely (1) explicit signaling from a set of possible line orientations and

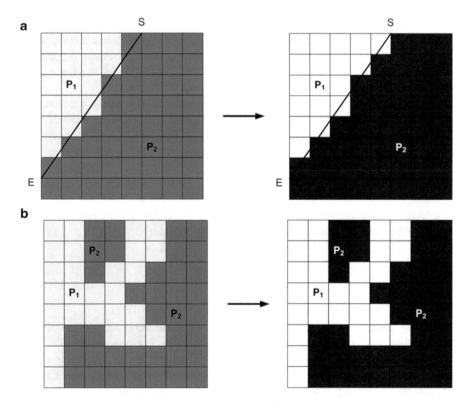

Fig. 6.6 Depth block partitioning into two regions, P_1 and P_2, using (**a**) Wedgelet and (**b**) contour type. The *left part* shows the original sample while the *right part* shows the resulting partition pattern [42]

Table 6.1 Intra-prediction modes in 3D-HEVC

Mode number	Name
0	INTRA_PLANAR
1	INTRA_DC
2–34	INTRA_ANGULAR
35	INTRA_DMM_WFULL
36	INTRA_DMM_CPREDTEX

positions. This can be implemented as a lookup table in which only the index needs to be transmitted. (2) The line can be predicted from already encoded blocks. The edges in neighboring blocks would normally continue in the current block and thus one or both of the end points of the dividing line can be obtained from the neighbors without any signaling. (3) Obtain the dividing line or contour indirectly through the co-located texture block, since the texture and the depth maps are correlated as they are both capturing the same scene.

One of the depth modeling types together with the signaling information that includes the partitioning data and their values is selected during the depth encoding process for blocks containing edges. For the other block containing homogeneous areas, normal intra-coding mode is used [24]. Table 6.1 gives the values of the intra-prediction modes in 3D-HEVC. The DMMs are defined with mode numbers 35 and 36 [43].

The DMM modes introduce extra computations during the mode decision process as the different DMM possibilities are tested [44]. This added complexity must be reduced with minimal effect on the coding efficiency such that these modes can be adopted in low-latency video applications.

The work in [26] proposed a solution to speed up the Wedgelet search process. The best coding mode for the depth data is done in four steps. A subset of all the intra-modes is first found by applying the sum of absolute transform difference (SATD) in rough mode decision. The SATD modes aid the search because the texture features are highly correlated with the line of the Wedgelet partition. All the Wedgelet patterns are classified into subset according to the angular modes, and the angular mode with the smallest SATD is obtained from the texture before testing the DMMs. The subset is then enriched by including the most probable modes from the neighboring blocks. Following this, the four DMM modes are searched in the corresponding angle subset and the one offering the smallest rate-distortion cost is selected as the optimal depth intra-prediction. The results in [26] show an encoding time saving of 60 % in depth intra-coding.

For most sequences, DMMs are selected only for a few blocks, that is, where there are edges. Thus, the work in [45] proposed a method to skip unnecessary rate-distortion cost calculations of DMMs, and thus achieve faster intra-coding. When the first mode in the full rate-distortion cost list is planar mode, there is high probability that the coding unit is flat or smooth. Thus, the DMM rate-distortion calculations can be skipped. Results presented in [45] show that the solution has less than 0.5 % miss rate, where a DMM should have been selected, with an encoding speedup of 27.8 %.

The work in [43] proposed a solution based on edge classification in the Hadamard transform domain. A set of prediction units that contain only straight horizontal or vertical edges is obtained. Results in [43] show that a large number of prediction units fall in this set, around 84 % on average; hence this can be exploited to obtain a fast algorithm. The probability that DMM is selected when the prediction unit lies within this set is very low and therefore its process can be omitted with very little loss in coding efficiency. On the other hand, if the prediction unit is not in the set, the probability that DMM is selected as the best mode increases and thus the DMMs should be tested. The results presented in [43] show an average improvement in mode decision time of 22.19 %.

The work in [46] uses an early termination criterion to speed up the mode decision. This criterion is based on the header rate of the DMM, which can be computed without performing the full encoding and decoding processes. The header rate is obtained from the context information of the neighbor prediction units that are needed for the arithmetic coding. The header rates are calculated both with and without segment-wise DC coding (SDC). The early termination is based on [46]

$$R_H(m) \leq \frac{J^{HEVC}}{\lambda}$$

where $R_H(m)$ is the header rate, J^{HEVC} is the minimum rate-distortion cost, λ is the Lagrange multiplier, and $m \geq 35$. When the DMM satisfies this equation it must be considered as a potential candidate in the rate-distortion cost calculations; otherwise it can be omitted. The results presented in [46] show an average encoding speedup of 33.89 %.

The Wedgelet patterns generated at initialization in each list can vary between 86 patterns for a 4×4 depth block and 1503 patterns for the 32×32 block. This set is then reduced through inter-component prediction using the co-located texture luma block (CTLB). The reduced set still has a large number of candidates. The works in [25] and [47] propose an inter-component depth modeling method that reduces the number of patterns to just six. The depth edges are found by employing a 1-D filter to each side of the CTLB. Before applying the filtering the resolution is considered, where (a) 4×4 and 8×8 depth blocks are doubled and linear interpolation is used to fill the missing data, (b) 16×16 depth blocks are unprocessed, and (c) 32×32 depth blocks are sampled by two by removing the odd rows and columns.

The start and end points calculated after filtering define the orientation of the Wedgelet. The set of six Wedgelet orientations used in [47] are shown in Fig. 6.7. The sum of absolute differences is used for the rate-distortion cost calculation to determine the best Wedgelet to signal. This technique only requires the signaling of the start and end coordinates in the bit stream and avoids the generation of the large lists since only one Wedgelet needs to be generated at the decoder. Results in [47] show an average of 4.7 % and 1.7 % improvement in encoding times for the hierarchical and all intra-prediction structures, respectively.

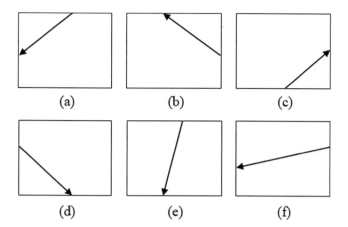

Fig. 6.7 The six Wedgelet orientations used in [47]

The inter-component Wedgelet mode (DMM mode 3) is not present in the latest version of the standard; however, it might be re-instated in future versions. The current standard incorporates the explicit Wedgelet signaling (DMM mode 1), which does not predict the partitions, but searches in a set of partitions at the encoder and inter-component contour mode (DMM mode 4) that employs inter-component prediction on the texture reference block for contour partitioning.

6.3.2   Depth Coding Based on Directional Intra-Coding and Flexible Block Partitioning

The predictive depth coding (PDC) algorithm [40, 41] has been recently proposed as an efficient solution for intra-coding of depth maps using an alternative paradigm, which combines the traditional directional prediction modes with a highly flexible block-partitioning scheme. Reported experiments have shown that PDC is able to achieve a higher rate-distortion performance than the current state-of-the-art 3D-HEVC encoder for depth map intra-coding. In this section, the whole PDC algorithm is described and some results comparing PDC with 3D-HEVC performance are presented.

6.3.2.1   Algorithm Overview

At its core, PDC algorithm uses a block-based hybrid coding approach based on intra-prediction and residue coding. The depth map is partitioned into non-overlapping blocks of 64×64 pixels that can be further partitioned into smaller

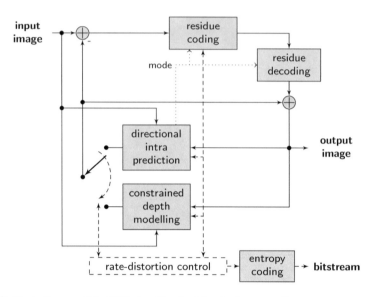

Fig. 6.8 Block diagram of the PDC algorithm for intra-depth map coding

sub-blocks using a flexible partitioning scheme. Each sub-block can be predicted using directional prediction or, alternatively, a new constrained depth modeling mode (CDMM), as illustrated in Fig. 6.8.

The used directional intra-prediction method is based on one of the HEVC standards [2]. However, PDC proposes significant improvements including adaptive mode pruning and efficient mode signaling. The block may alternatively be encoded using CDMM, designed for explicitly signaling the edges that are difficult to predict. This kind of edges are typically observed in the bottom-right region of the block, which cannot be predicted by directional prediction using left and top neighboring block samples. CDMM allows to explicitly signal an approximation of the edges in the block and surrounding smooth areas.

PDC does not use transform-based residual information coding. Instead, the residual information is encoded using a straightforward method that applies linear approximations to the residue signal, depending on the chosen prediction mode. The approximation linear coefficients are transmitted using a depth lookup table (DLT), as proposed in 3D-HEVC.

On the encoder side, most of the possible combinations of block partitioning and coding modes are examined and the best one is selected according to a Lagrangian rate-distortion cost. Context-adaptive m-ary arithmetic coding (CAAC) [48] is used for entropy coding, including the symbols that represent the flexible block partition, directional prediction modes, constrained depth modeling mode, and residue coding.

Fig. 6.9 Possible block sizes in PDC and respective label numbers

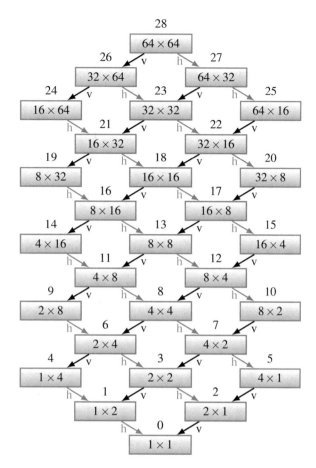

6.3.2.2   Flexible Block Partitioning

In PDC, each block may be partitioned using a flexible scheme based on bintree and quadtree methods. Bintree partitioning recursively divides the input block, either in the vertical or horizontal directions, down to the 1×1 size, as illustrated in Fig. 6.9. Some block sizes with a high ratio between horizontal and vertical dimensions (ratios larger than 4, e.g., 64×1) are not used, because they have a small impact on the algorithm performance and, thus, computational complexity can be reduced.

The bintree block-partitioning scheme is combined with quadtree partitioning, in order to further reduce the encoder's computational complexity. Three quadtree levels are defined at block sizes 16×16, 32×32, and 64×64. The four partitions, generated by each quadtree partitioning, are processed using a raster scan order. For each available quadtree level, the bintree partitioning is used down to a predefined minimum block size.

6.3.2.3 Directional Intra-Prediction

When combined with the flexible block-partitioning scheme, the directional prediction framework provides an efficient representation of depth map edges. The directional intra-prediction includes the planar, DC, and 33 angular prediction modes. PDC further improves the prediction of depth map signals, by using predefined and adaptive reduction of directional modes, for minimal signaling.

As discussed before, PDC allows a large set of block sizes for prediction. Since some block sizes are very small or narrow, some directional intra-prediction modes (e.g., the ones with adjacent directions) may be redundant, resulting in similar prediction results. Thus, PDC uses a predefined reduced set of available directional modes, mainly for smaller block sizes, in order to save unnecessary calculations and overhead bits.

The adaptive reduction of directional modes is another improvement of PDC algorithm, which has similar coding advantages. It exploits the large amount of smooth areas present in depth maps. In these areas, various directional modes may produce the same predicted samples. To avoid this situation adaptive reduction of directional modes is used, based on the reference samples in the block neighborhood. Three groups of directional prediction modes are defined, depending on their reference samples. The modes of one group are disabled when the associated reference samples are exactly constant. The used groups of modes and associated reference samples, identified by the block neighboring region, are shown in Table 6.2. Group 1 contains all the directions that generate a prediction signal, exclusively based on the top and left block neighborhood including the top-left pixel. When these reference samples are constant, the associated modes of group 1 are disabled. DC mode can be chosen in place of the disabled modes of group 1, since it produces the same predicted samples. When the samples of the neighbor left and down-left regions are constant, the modes of group 2 can be disabled. In this case, the angular 10 mode (horizontal) is able to substitute these modes, producing the same results. Group 3 contains those modes that depend on top and top-right neighbor regions, and can be replaced by angular mode 26 (vertical).

6.3.2.4 Constrained Depth Modeling Mode

The main idea behind CDMM is to boost the intra-directional prediction, by providing an alternative method that explicitly encodes depth map edges that are difficult to predict. These edges are often present in the bottom-right region of the

Table 6.2 Groups of prediction modes defined according to the block neighbor regions

Group	Neighbor regions	Prediction modes
1	Top and left and top-left	Modes 10–26, planar
2	Left and down-left	Modes 2–9
3	Top and top-right	Modes 27–34

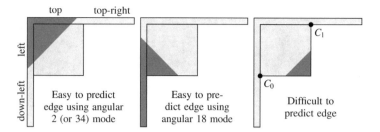

Fig. 6.10 Example of simple edge prediction (*left* and *middle*) and difficult edge prediction (*right*)

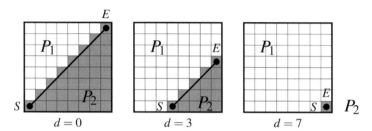

Fig. 6.11 Block partition examples using the proposed constrained depth modeling mode

block. Directional prediction framework reasonably predicts straight edges coming from the left or top block neighborhood. However, when an edge does not touch the left or top neighbor samples, like the one shown in the right block of Fig. 6.10, it becomes difficult to predict.

CDMM is inspired on the Wedgelet depth modeling mode used in 3D-HEVC, but several restrictions were applied to its design, in order to make it more efficient in the context of the PDC algorithm. CDMM divides the block into two partitions, which are approximated by constant values. The block partitioning occurs between two points of the right and bottom margins of the predicting block. As a second restriction, the line drawn between the two chosen points should be parallel to the anti-diagonal defined by the down-left and top-right block corners.

Figure 6.11 illustrates some partition options of the CDMM for an 8×8 square block. Due to the imposed constraints only one parameter is needed to signal the block partitioning in CDMM, being represented in Fig. 6.11 by the offset d. In this example, eight different partitions that vary between the minimum offset, $d = 0$, and the maximum offset, $d = 7$, can be employed.

The restriction on the block-partitioning slope is advantageous in terms of computational complexity because it avoids testing many block partitions with different slopes. Furthermore, by using a unique partition slope associated with the block size, no bit stream overhead is required for its transmission. The main disadvantage of this partitioning restriction is the reduced flexibility to approximate depth map edges. However, the proposed PDC algorithm is able to alleviate this issue by combining CDMM with the flexible partitioning scheme.

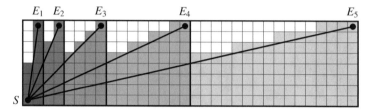

Fig. 6.12 CDMM partition slopes provided by flexible block partitioning

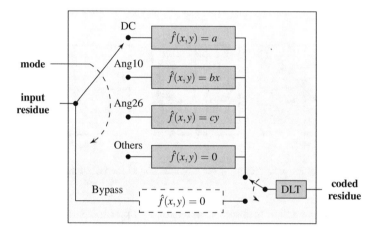

Fig. 6.13 Detailed diagram of the PDC residue coding method

The large amount of block sizes generated through flexible partitioning provides up to five different CDMM partitioning slopes according to the possible down-to-top diagonals. The illustrated overlapped block sizes in Fig. 6.12 represent all the block width/height ratios available in PDC.

CDMM block partitioning generates two partitions, whose depth values are approximated by using a constant value. For $P1$ partition, the approximation coefficient is derived from the block neighborhood. The constant approximation of $P2$ partition is explicitly transmitted to the decoder.

6.3.2.5 Residual Signal Coding

The flexible block-partitioning scheme combined with the directional intra-prediction and CDMM provides very efficient prediction results. For this reason, PDC does not use the transform-based coding, but an alternative approach based on linear modeling, for low computational complexity. Figure 6.13 illustrates the schematic of the proposed residue coding method. Four approximation models are available: constant, horizontal linear and vertical linear, as well as a special case of null residue. Depending on the chosen prediction mode, which is known in both the encoder and decoder, one of the available residue approximation models is transmitted. For a more efficient rate-distortion coding, PDC allows to bypass residue approximation, through the use of a binary flag. In order to better encode pre-quantized depth maps, the DLT algorithm is used to encode linear approximation coefficients.

6.3.2.6   Rate-Distortion Performance

The intra-coding performance of the PDC and 3D-HEVC algorithms is compared using the three-view configuration for the multi-view video coding with depth data scenario, as proposed by MPEG in common test condition document for 3D video core experiments [22]. 3D-HEVC results were produced using reference software version HTM-13.1, with all-intra configuration enabled, but the contour depth modeling mode, in order to prevent inter-component prediction.

For evaluation purposes, the methodology recommended by ISO/IEC and ITU-T JCT-3V group is used. The PSNR quality of the virtual views, based on the decoded depth data and the original texture views, is computed against the reference virtual views, based on the original uncompressed depth and original texture views. The average luminance PSNR quality of six intermediate views placed between the positions of the encoded depth maps is used. For the purpose of view synthesis, state-of-the-art view synthesis software for linear camera arrangement implemented in HTM software is used [49]. The sum of the bitrate used to encode the three depth map views is considered in the evaluation process.

Figure 6.14 presents the average Bjontegaard Delta Bitrate [50] results of PDC relative to 3D-HEVC, for eight different test sequences, under two configurations with different distortion metrics: the sum of square errors (SSE) and the view synthesis optimization (VSO). These results clearly show the advantage of the PDC algorithm over the state-of-the-art 3D-HEVC approach, either using SSE or VSO

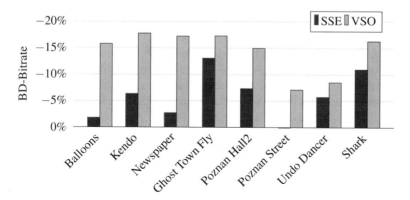

Fig. 6.14 Average BD-BR values of PDC relative to 3D-HEVC using SSE and VSO distortion metrics

distortion metrics. The average bitrate savings of PDC over 3D-HEVC using all-intra configuration and SSE distortion metric is approximately 6 %. The average PDC gain is superior when VSO method is used in both algorithms, achieving 14.3 % of bitrate reduction for the same quality of virtual views.

6.3.3   Down-Sampling and Post-processing Depth Map Techniques

The depth maps may contain large homogeneous areas making down-sampling of the data attractive as it allows saving of bandwidth. The down-sampled depth is then restored back using up-sampling at the receiver before view synthesis, such that the texture and the depth videos are at the same resolution. However, this process can distort the sharp edges in the depth maps introducing artifacts in the synthesized views.

The easiest way to down-sample a frame is through the pixel-dropping scheme, where to, for example down-sample by half, neighboring pixels of each even numbered row and column are removed. This technique is the least computational demanding and is implemented using the equation below [51]:

$$\text{Depth}_{\text{down}}(x, y) = \text{Depth}_{\text{orig}}(x * \text{step}, \ y * \text{step})$$

where $\text{Depth}_{\text{orig}}$ is the original depth image, $\text{Depth}_{\text{down}}$ is the down-sampled depth image, x and y represent the pixel coordinate, and step is the down-sampling factor.

The down-sampled depth maps are modified in [52] through filtering techniques to help the up-sampling stage. Other improvements to the down-sampling methods are found in [53] where adaptive depth filtering is done prior to down-sampling to increase the uniformity in the adjacent regions.

At the receiver, the down-sampled depth maps need to be restored to their original resolution. The most commonly used methods are nearest-neighbor and bicubic methods followed by color-based and edge-based solutions that exploit the texture information, which is also available at the receiver. The nearest-neighbor algorithm duplicates pixels in the horizontal and vertical directions, where the number of duplications of each pixel depends on the original scaling factor. This method is fast but results in a checkerboard effect, which becomes more pronounced as the scaling factor increases [54]. This will generate errors close to the edges, as can be seen in Fig. 6.15. These errors will translate into artifacts in the synthesized views.

The bicubic image interpolation scheme considers square areas of 16 pixels and is more complex than the nearest neighbor. This method is given by

$$p(x, y) = \sum_{i=0}^{3}\sum_{j=0}^{3} a_{ij} x^i y^j$$

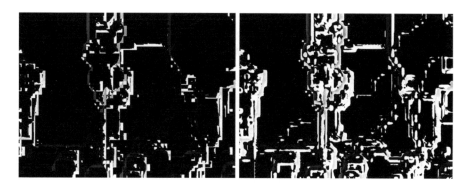

Fig. 6.15 Difference between original depth map and the up-sampled depth map using the nearest-neighbor interpolation (*left*) or bicubic interpolation method (*right*) algorithm for the *Balloons* sequence. In this case the up-sampling factor is 16 and *black pixels* indicate no difference [55]

where a_{ij} represents the 16 coefficients. The resulting image is generally smoother than the nearest-neighbor technique [53]. Figure 6.15 shows the difference between the original depth map and the up-sampled depth map using the bicubic interpolation technique.

The color-based method used in [51] assumes that similar color pixels in neighboring regions in the texture image will have the same depth value in the corresponding depth map. In this technique, the up-sampled depth map and the texture image are first aligned, by using the known points corresponding to the down-sampled image. The missing pixels are then filled by considering the four known neighbor pixels and applying a weight depending on the color distance. This method produces better results than the previous techniques, at the expense of an increase in complexity.

Edge-based techniques try to improve the edge information in the depth map through the knowledge that edges in the depth map must be present in the corresponding texture image. The method in [56] uses the nearest-neighbor interpolation followed by filtering. The Canny edge detector is applied to the texture data followed by a morphological operator to establish, from the down-sampled map, where the edges of interest should lie. Once the results from the Canny edge detector are cleaned, the edge contours are found. These are dilated and two contours are created, one classified as foreground and the other as background. For both contours, dilation and removal of the original line are iteratively performed to generate a number of layer levels. Given that the depth pixels and the edges should have the same values, the pixels are then painted accordingly. A 7×7 block is considered to paint all the pixels on the other edge layers.

Other edge-based up-sampling methods include [57], where an edge-directed interpolation scheme is used. The edges are found in the texture images, giving more reliable edge information. Furthermore, Yang [52] use an edge-based weighting function to preserve the edges. Similar to the color-based techniques, these methods increase the complexity of the system. Moreover, since the texture camera view and the depth map view might be slightly misaligned in practice, some errors at the edges will still be present introducing artifacts in the synthesized views.

Post-processing filters have also been proposed to reduce the errors generated in depth maps by up-sampling methods. The method in [58] proposes a post-processing method after nearest-neighboring up-sampling based on three filters, specifically a 2D median filter, a frequent-low-high reconstruction filter, and a bilateral filter. While median filter smooth blocking artifacts, the frequent-low-high filter is used to recover object boundaries. The bilateral filter is used to remove the remaining errors after the previous two filters. In [59], a dilation filter is used to improve depth map after up-sampling with linear interpolation. An analysis of these filters in [60] has shown that dilation filter tends to be more effective when used in-loop by view synthesis prediction (VSP) techniques for texture video coding, while the method of [58] presents better performance in the post-processing stage [61].

6.4 Conclusion

This chapter covered the need for the transmission of depth data as a means to reduce bandwidth requirements for multi-view video applications. We saw that encoding this content is not optimal when using traditional video encoding schemes that are designed for the texture video and thus new tools are needed. The current state-of-the-art and new techniques were discussed giving a good insight into the topic and providing valuable references to the reader.

More research work is needed in 3D video coding, both in terms of improving coding efficiency and reducing encoding and decoding times. Depth map coding will play an important role in providing a more immersive 3D video experience in future applications and services. Thus, better ways of preserving the correctness of the characteristics of depth maps and their efficient encoding when transmitted over error-prone channels will remain a hot topic in the coming years.

Acknowledgements The authors would like to thank the Interactive Visual Media group at Microsoft Research for providing the *Ballet and Breakdancers* test sequences, Nagoya University for providing the *Balloons* test sequence, and National Institute of Information and Communications Technology (NICT) for providing the *Sharks* test sequence for research purposes.

References

1. (2010) Report on experimental framework for 3D video coding. ISO/IEC JTC1/SC29/WG11, N11631
2. ITU-T and ISO/IEC JTC 1/SC 29 (MPEG) (2013) High efficiency video coding, Recommendation ITU-T H.265 and ISO/IEC 23008-2
3. Sullivan G, Ohm J, Han W, Wiegand T (2012) Overview of the high efficiency video coding (HEVC) standard. IEEE Trans Circ Syst Video Technol 22(12):1649–1668

4. ITU-T and ISO/IEC JTC1 (2010) Advanced video coding for generic audiovisual services, ITU-T Recommendation H.264 and ISO/IEC 14496-10 (MPEG-4 AVC)
5. Müller K, Schwarz H, Marpe D, Bartnik C, Bosse S, Brust H, Hinz T, Lakshman H, Merkle P, Rhee F, Tech G, Winken M, Wiegand T (2013) 3D high-efficiency video coding for multi-view video and depth data. IEEE Trans Image Process 22(9):3366–3378
6. Schwarz H, Bartnik C, Bosse S, Brust H, Hinz T, Lakshman H, Merkle P, Müller K, Rhee H, Tech G, Winken M, Marpe D, Wiegand T (2012) Extension of high efficiency video coding (HEVC) for multiview video and depth data. In: Proceedings of the 19th IEEE international conference on image processing. pp 205–208
7. Sullivan G, Boyce J, Chen Y, Ohm J, Segall C, Vetro A (2013) Standardized extensions of high efficiency video coding (HEVC). IEEE J Selected Topics Signal Process 7(6):1001–1016
8. Tanimoto M, Tehrani M, Fujii T, Yendo T (2011) Free-viewpoint TV—a review of the ultimate 3DTV and its related technologies. IEEE signal processing magazine. pp 67–76
9. Do L, Zinger S, Morvan Y, With P (2009) Quality improving techniques in DIBR for free-viewpoint video. In: Proceedings of 3DTV conference: the true vision—capture, transmission and display of 3D video
10. Müller K, Smolic A, Dix K, Kauff P, Wiegand T (2010) Reliability-based generation and view synthesis in layered depth video. In Proceedings of the IEEE 10th workshop on multimedia signal processing. pp 34–39
11. Li W, Zhou J, Li B, Sezan I (2009) Virtual view specification and synthesis for free viewpoint television. IEEE Trans Circ Syst Video Technol 19(4):533–546
12. Shoemake K (1985) Animation rotation with quaternion curves. In: Proceedings of 12th annual conference on computer graphics and interactive techniques, vol 19, no. 3, pp 245–254
13. Lee C, Ho Y (2009) View synthesis using depth map for 3D video. In: Proceedings of the Asia-Pacific Signal Information Processing Association annual summit and conference. pp 350–357
14. Oh K, Yea S, Vetro A, Ho Y (2010) Virtual view synthesis method and self-evaluation metrics for free viewpoint television and 3D video. Int J Imaging Syst Technol 20(4):378–390
15. Yang X, Lui J, Sun J, Li X, Liu W, Gao Y (2011) DIBR based view synthesis for free-viewpoint television. In: Proceedings of 3DTV conference: the true vision—capture, transmission and display of 3D video
16. Mori Y, Fukushima N, Fujii N, Tanimoto M (2008) View generation with 3D warping using depth information for FTV. In: Proceedings of 3DTV conference: the true vision—capture, transmission and display of 3D video
17. Zarb T, Debono, C (2014) Depth-based image processing for 3D video rendering applications. In: Proceedings of the 21st international conference on systems, signals and image processing. pp 215–218
18. Tran A, Harada, K (2013) View synthesis with depth information based on graph cuts for FTV. In: Proceedings of the 19th Korea-Japan joint workshop on frontiers of computer vision. pp 289–294
19. Xu J, Yan F, Cao X (2014) Stereoacuity-guided depth image based rendering. In: Proceedings of the IEEE international conference on multimedia and expo
20. Lei J, Zhang C, Fang Y, Gu Z, Ling N, Hou C (2015) Depth sensation enhancement for multiple virtual view rendering. IEEE Trans Multimedia 17(4):457–469
21. JCT3V-J1005 (2014) 3D-HEVC test model 10, joint collaborative team on 3D video coding extension development of ITU-T SG 16 WP 3 and ISO/IEC JTC 1/SC 29/WG 11
22. JCT3V-G1100 (2014) Common test conditions of 3DV core experiments. Joint collaborative team on 3D video coding extension development of ITU-T SG 16 WP 3 and ISO/IEC JTC 1/SC 29/WG 11
23. Jager F (2012) Simplified depth map intra coding with an optional depth lookup table. In: Proceedings of the international conference on 3D imaging. pp 1–4
24. Merkle P, Bartnik C, Müller K, Marpe D, Wiegand T (2012) 3D video: depth coding based on inter-component prediction of block partitions. In: Proceedings of the picture coding symposium. pp 149–152

25. Song Y, Ho Y-S (2013) Simplified inter-component depth modeling in 3D-HEVC. In: Proceedings of the IEEE 11th image, video and multidimensional signal processing workshop
26. Zhang M, Zhao C, Xu J, Bai H (2013) A Fast depth-map wedgelet partitioning scheme for intra prediction in 3D video coding. In: Proceedings of the IEEE international symposium on circuits and systems. pp 2852–2855
27. Merkle P, Muller K, Marpe D, Wiegand T (2015) Depth intra coding for 3D video based on geometric primitives. IEEE Trans Circ Syst Video Technol
28. Zamarin M, Salmistraro M, Forchhammer S, Ortega A (2013) Edge-preserving intra depth coding based on context-coding and H.264/AVC. Multimedia and expo (ICME), 2013 IEEE international conference on, pp 1–6
29. Oh B, Wey H, Park D (2012) Plane segmentation based intra prediction for depth map coding. In: Proceedings of the picture coding symposium. pp 41–44
30. Shen G, Kim W-S, Ortega A, Lee J, Wey H (2010) Edge-aware intra prediction for depth-map coding. In: Proceedings of the 2010 IEEE international conference on image processing. pp 3393–3396
31. Krishnamurthy R, Chai B, Tao H, Sethuraman S (2001) Compression and transmission of depth maps for image-based rendering. In: Proceedings of the 2001 IEEE international conference on image processing. pp 828–831
32. Merkle P, Morvan Y, Smolic A, Farin D, Müller K, With P, Wiegand T (2009) The effects of multiview depth video compression on multiview rendering. Image Commun 24(1–2):73–88
33. Zia W, Diepold K, Sarkis M (2010) Fast depth map compression and meshing with compressed tritree. In: Proceedings of the 9th Asian conference on computer vision—volume part II. pp 44–55
34. Shen G, Kim W-S, Narang S, Ortega A, Lee J, Wey H (2010) Edge-adaptive transforms for efficient depth map coding. In: Proceedings of the picture coding symposium
35. Hu W, Cheung G, Ortega A, Au O (2015) Multiresolution graph Fourier transform for compression of piecewise smooth images. IEEE Trans Image Process 24(1):419–433
36. Graziosi D, Rodrigues N, Pagliari C, Silva E, Faria S, Perez M, Carvalho M (2010) Multiscale recurrent pattern matching approach for depth map coding. In: Proceedings of the picture coding symposium. pp 294–297
37. Graziosi D, Rodrigues N, Pagliari C, Faria S, Silva E, Carvalho M (2010) Compressing depth maps using multiscale recurrent pattern image coding. Electron Lett 46(5):340–341
38. Francisco N, Rodrigues N, Silva E, Carvalho M, Faria S, Silva V, Reis M (2008) Multiscale recurrent pattern image coding with a flexible partition scheme. In: Proceedings of the 15th IEEE international conference on image processing
39. Lucas L, Rodrigues N, Pagliari C, Silva E, Faria S (2012) Efficient depth map coding using linear residue approximation and a flexible prediction framework. In: Proceedings of the 19th IEEE international conference on image processing
40. Lucas L, Rodrigues N, Pagliari C, Silva E, Faria S (2013) Predictive depth map coding for efficient virtual view synthesis. In: Proceedings of the 20th IEEE international conference on image processing
41. Lucas L, Wegner K, Rodrigues N, Pagliari C, Silva E, Faria S (2015) Intra predictive depth map coding using flexible block partitioning. IEEE Trans Image Process 24(11):4055–4068
42. Müller K, Merkle P, Tech G, Wiegand T (2012) 3D video coding with depth modeling modes and view synthesis optimization. In: Proceedings of the signal and information processing association annual summit and conference
43. Park C-S (2015) Edge-based intramode selection for depth-map coding in 3d-HEVC. IEEE Trans Image Process 24(1):155–162
44. Chen Y, Tech G, Wegner K, Yea S (2014) Test model 9 of 3D-HEVC and MV-HEVC, document JCT3V-I1003, ITU-T SG 16 WP3 and ISO/IEC JTC 1/Sc 29/WG 11
45. Gu Z, Zheng J, Ling N, Zhang P (2013) Fast depth modeling mode selection for 3D HEVC depth intra coding. In: Proceedings of the IEEE international conference on multimedia and expo workshops

46. Park C-S (2015) Efficient intra-mode decision algorithm skipping unnecessary depth-modelling modes in 3D-HEVC. Electronics Lett 51(10):756–758
47. Ricci K, Debono C (2015) A fast inter-component depth modeling technique for 3D high efficiency video coding. In: Proceedings of 3DTV-conference 2015—immersive and interactive 3D media experience over networks
48. Witten I, Neal R, Cleary J (1987) Arithmetic coding for data compression. Commun ACM 30(6):520–540
49. ISO/IEC JTC1/SC29/WG11 MPEG2013/M31520, Wegner K, Stankiewicz O, Tanimoto M, Domanski M (2013) Enhanced view synthesis reference software (VSRS) for free-viewpoint television
50. Bjøntegaard G (2001) Calculation of average PSNR differences between RD-curves. ITU-T SG 16 Q.6 VCEG, Doc. VCEG-M33
51. Wildeboer M, Yendo T, Tehrani M, Fujii T, Tanimoto M (2010) Color based depth up-sampling for depth compression. In: Proceedings of the picture coding symposium. pp 170–173
52. Yang Y, Zheng J (2013) Edge-guided depth map resampling for HEVC 3D video coding. In: Proceedings of the international conference on virtual reality and visualization. pp 132–137
53. Schwarz S, Olsson R, Sjöström M, Tourancheau S (2012) Adaptive depth filtering for HEVC 3D video coding. In: Proceedings of the picture coding symposium. pp 49–52
54. Gonzalez R, Woods R (2007) Digital image processing, 3rd edn. Prentice Hall
55. Zammit L (2015) Depth map processing for down-sampled MV-HEVC encoded depth video. M.Sc. Dissertation, University of Malta
56. Graziosi D, Dong T, Vetro A (2012) Depth map up-sampling based on edge layers. In: Proceedings of the Asia-Pacific signal information processing association annual summit and conference
57. Huiping D, Li Y, Jinbo Q, Juntao Z (2012) A joint texture/depth edge-directed up-sampling algorithm for depth map coding. In: Proceedings of the IEEE international conference on multimedia and expo. pp 646–650
58. Oh K-J, Yea S, Vetro A, Ho Y-S (2009) Depth reconstruction filter and down/up sampling for depth coding in 3-D video. IEEE Signal Process Lett 16:747–750
59. Lee S, Lee S, Wey H, Lee J (2012) 3D-AVC—CE6 related results on Samsung's in-loop depth resampling. ISO/IEC JTC1/SC29/WG11 Document M23661, San Jose, USA
60. Graziosi D, Rodrigues N, Silva E, Carvalho M, Faria S, Tian D, Vetro A (2013) Analysis of depth map resampling filters for depth-based 3D video coding. Conference on telecommunications—ConfTele
61. Guarda A, Santos J, Graziosi D, Rodrigues N, Faria S (2014) A novel trilateral filter technique for depth map processing in 3D video coding. In: Proceedings of the 3DTV conference—3DTV-CON, Budapest, Hungary

Chapter 7
Hybrid Broadcast Broadband for the Delivery of 3D Video

Asimakis Lykourgiotis, Tasos Dagiuklas, Ilias Politis, Hugo Marques, Jonathan Rodríguez, and Hélio Silva

Abstract Fuelled by the need to increase their market share, broadcast service providers enhance their services with media and social networking. Hybrid broadcast broadband (HBB) television (HbbTV) has emerged as a promising means of content distribution for enriched and interactive television services. Specifically, the convergence of terrestrial digital video broadcasting (DVB) services and peer-to-peer (P2P) overlay networks is an up-and-coming telecommunication design. In this chapter, we discuss how this vision came into fruition, within the context of the European-funded project ROMEO. ROMEO project aims at delivering three-dimensional (3D) multi-view content to fixed, mobile and portable users with guaranteed minimum quality of experience (QoE). In particular, we elaborate on the P2P overlay construction process as well as on techniques to optimise its performance. Moreover, we present a packetisation scheme that facilitates the task of synchronising the streams that are being received from DVB and P2P networks. Additionally, ROMEO architecture supports mobility by exploiting the evolved packet core design and enhancing content delivery, particularly for mobile devices. Finally, we specify the life cycle of the ROMEO content from a QoE perspective from core network to access network including scenarios involving multiple internet service providers.

A. Lykourgiotis (✉) • I. Politis
Hellenic Open University, Parodos Aristotelous 18, Patras 26335, Greece
e-mail: alykourgiotis@eap.gr; ilpolitis@eap.gr

T. Dagiuklas
London South Bank University, 103 Borough Road, London, SE1 0AA, UK
e-mail: tdagiuklas@lsbu.ac.uk

H. Marques • J. Rodríguez • H. Silva
Instituto de Telecomunicações, Campus Universitário Santiago, Aveiro, Portugal
e-mail: hugo.marques@av.it.pt; jonathan@av.it.pt; heliosilva@av.it.pt

© Springer Science+Business Media New York 2017   167
A. Kondoz, T. Dagiuklas (eds.), *Connected Media in the Future Internet Era*,
DOI 10.1007/978-1-4939-4026-4_7

7.1 Introduction

We are moving to a new era of convergence between media consumer and content provider. This shift is due to the new methods of media consumption via sharing using technologies such as converged heterogeneous networks, new transport methods (e.g. peer-to-peer—P2P, multiple multicast trees—MMT, etc.), personal/user generated tools and social media [15]. As our lifestyles become more connected, even the passive behaviour of watching television is turning into a very active process, where viewers are multitasking on their mobile devices while watching television. This shift poses new challenges in jointly optimising media networking and sharing personalised content. Social TV refers to the convergence between broadcasting, media networking and social networking. This gives the capability to the end-users to communicate, publish and share content among themselves and interact with the TV program. Consumers have aggressively adopted online video services (e.g. Netflix, YouTube, etc.). As more providers, more content and more devices become available, consumers seem ready to take full advantage. More consumers also expect to see increased use of video on laptops, tablets and smartphones than on any other devices [5].

Looking from users' perspective, uploading their own content while enjoying the commonly shared content, as well as interacting either with the content which is being displayed or with their colleagues in the socially collaborating group, is an important parameter to make the whole social interaction truly enjoyable. A user on a mobile device (e.g. PDA, Tablet, etc.) should be able to access the content from the closest point (distance, signal strength and availability) where the service is available. This particular service point is expected to also provide media processing capabilities to match the needs of the mobile terminal by pre-processing the available content in accordance to the given device capabilities, e.g. processing capability, resolution, content format it supports, etc. This could be supported by media storage and processing clouds that either the users will subscribe to (as an additional service) or their mobile network operator will provide it (within the subscription package) for increased user QoE and hence increase the network's customer base. The TV experience is enhanced by additional services.

As demonstrated by the introduction of HDTV through satellite and digital terrestrial television , broadcast TV networks have pioneered the transition to high quality TV services and consistently play a leading role in the introduction of new TV standards. However, broadcast networks do not benefit from the native return path, which is required to deliver advanced TV services. Broadband operators can deliver on-demand content and full-scale interactivity but their geographic reach for TV services is limited. A number of standards, which can be natively implemented in connected TVs or stand-alone devices, are also available for hybrid broadcast broadband (HBB) TV integration [7]. HbbTV combines the advantages of broadband for delivering individual choice of on-demand content with the efficiency of broadcasting (satellite, digital terrestrial) for making high quality TV simultaneously available to a large audience. Hybrid TV solutions also require very short time-to-market implementation of linear and non-linear TV services for consumers. The European Commission funded integrated research project ROMEO [18] introduces a novel architecture for HBB networks in order to support 3D multi-view content [14]. In this chapter, the architecture and main functional entities of the ROMEO platform are being presented, along with a discussion of the life cycle of the multimedia content.

The remainder of this chapter is as follows. Initially, in Sect. 7.2 we present an overview of the ROMEO concept, briefly describing the architecture and the basic supported use cases of the ROMEO platform. Subsequently, in Sect. 7.3, we elaborate on the different modules that compose the server and the peers and enable the timely 3D content delivery and consumption. Thereafter, in Sect. 7.4, we describe how the 3D video content is generated as well as the content packetisation scheme that enables synchronisation between the broadcast and the broadband access network. Afterwards, in Sect. 7.5, we elaborate on the life cycle of the content from a QoE perspective from the core network to the access network and the fixed and mobile users. Finally, in Sect. 7.6, we conclude the chapter.

7.2 Hybrid Architecture: The ROMEO Concept

ROMEO project aims at delivering 3D multi-view content to fixed, mobile and portable users with guaranteed minimum QoE. To achieve this, DVB [8] technology is combined with the P2P Internet technology to take advantage of both DVB and Internet Protocol-based (IP) networks. DVB provides a reliable and deterministic content delivery path and by exploiting this characteristic of DVB, ROMEO users retrieve the content with a quality of DVB stream in the worst case. On the other hand, IP network allows delivering multiple high quality views, which cannot be delivered via DVB due to its bandwidth limitations. As a result, the broadcaster can deliver high quality stereoscopic 3D content to end-users via DVB and at the same time stream a set of supplementary 3D multi-view content (e.g. additional viewpoints with their respective spatial audio), through the P2P network. ROMEO also exploits and augments the evolved packet core (EPC) [16] functionalities to deliver 3D multi-view content to mobile users. As another capability, the remote users are provided with a real-time audio-visual (A/V) communication channel so that they can share their experiences while watching the high quality 3D media. These collaborating users can also share their own content between each other.

In Fig. 7.1, an overview of ROMEO services is provided. As can be seen, a user is connected to the ROMEO system as a member of P2P topology (MMT). ROMEO topology is managed by the central ROMEO server, so the user must be connected to the server as the first step to join the ROMEO system. ROMEO server is also responsible for distributing the P2P streams corresponding to multiple view 3D video and audio to the whole ROMEO system via super-peers. Delivered streams are then delivered through the MMT [3] and peers receive these streams from their

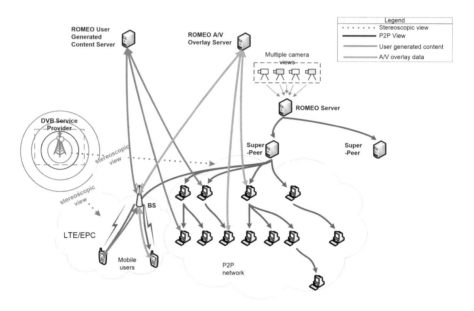

Fig. 7.1 ROMEO architecture

parents. ROMEO P2P streams are shown as violet in Fig. 7.1. It is assumed that each user will receive main stereoscopic view from DVB, which is shown as beige in Fig. 7.1. This stereoscopic view is then combined and synchronised with the additional views received from P2P to provide users high quality 3D multi-view experience. The assumption of DVB availability for each of the users guarantees the seamless content consumption in case of deteriorations in the network conditions.

During the consumption of the live stream, ROMEO users can collaborate and share their own content as mentioned before. In order to minimise the impact of these services (such as transfer of the user generated content—UGC) on the P2P live streaming media during the collaboration, UGC (shown as orange in Fig. 7.1) and A/V overlay (shown as green in Fig. 7.1) data paths are separated from the P2P delivery path with an independent client–server-based architecture.

The incorporation of DVB and IP service providers at the same time requires the coordination between the content received from these two providers. An operator having both DVB- and IP-based content services is an ideal business model for providing the ROMEO services which may not be the case for most of the time. As an alternative business model, virtual media operators would emerge as a new entity that will provide media services to end-users by contracting with companies providing broadcast and broadband services. With the development of ROMEO and similar hybrid media delivery platforms, these virtual operators may come into play to provide services exploiting hybrid delivery technologies in near future.

To sum up, the ROMEO system mainly involves two use cases. The first use case scenario is the consumption of 3D audio/video over hybrid delivery paths. In this scenario, the hybrid network is delivering a 3D live stream using either only one of the available paths or a combination of them (i.e. DVB and P2P over IP) at a given time to maximise the level of QoE offered to the user, including adaptation in response to varying network conditions, triggering different media delivery options (adjusting number of views, quality of views, etc.). The second use case scenario incorporates the collaborating group communication and sharing of UGC. In this scenario, the hybrid network is providing a real-time audio-visual communication channel to connect collaborating users while they are receiving the live stream via their DVB and/or P2P connections. In this scope, users can initiate a collaborating group and specify the members of the group. Users can generate or use existing content and share it with the members of the collaborating group for further social interaction.

7.3   A Closer Look at ROMEO Building Blocks

The ROMEO approach for network delivery of synchronised 3D media networking with real-time audio-visual communication support involves a variety of building blocks. Initially, there is the main server, where the ROMEO 3D media content is originated and stored. It belongs to a specific DNS domain (typically associated with the service brand) and it is property of the service owner. Additionally, UGC server is responsible for storing and managing UGC, that is, content created by ROMEO users. Moreover, there is the multiple coding server, which converts UGC to multiple file formats for compatibility and quality of experience (QoE) reasons. The A/V overlay server provides a real-time audio-visual communication channel which connects all collaborating users. Furthermore, there are the super-peers, that are proxies/replicas of the main server and are placed at the premises of every internet service provider (ISP) that has an agreement with the service owner. The super-peers are responsible to serve peers from a specific geographical area or ISP. For large scale deployments, ROMEO super-peer nodes, which are also shown in Fig. 7.1, behave as the replicas of the ROMEO server. Lastly, there are the peers, which are fixed, portable or mobile devices belonging to the ROMEO end-users, who will consume the ROMEO 3D media content.

Figure 7.2 shows the ROMEO enabler components which provide network agnostic behaviour to the ROMEO system. When a user wants to join the ROMEO system, its device and network capabilities are queried by the network monitoring subsystem (NMS) component (Sect. 7.3.2.5). The topology builder (TB) component inserts the peer in the most suitable position in the MMT topology by taking these network monitoring reports into consideration (Sect. 7.3.1.1). During run-time, the NMS peer module periodically reports the conditions of each peer to the multicast tree manager (MTM) server module, presented in Sect. 7.3.1.2, which is intrinsically related to the TB. By using the updated peer records and a complex algorithm, the TB decides whether it should promote or demote peers from their current position on each tree. Chunk selection (CS), described in Sect. 7.3.2.3, and media aware proxy (MAP), defined in Sect. 7.5.3, will perform network adaptation for the peers in case

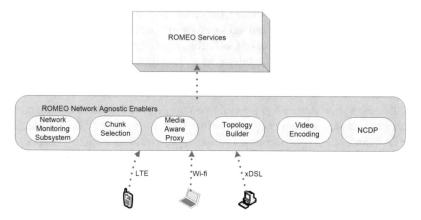

Fig. 7.2 ROMEO heterogeneous networks solution

of having insufficient bandwidth. In this way, it is aimed to provide service to each peer according to its device and network capabilities regardless of its network access method. This capability requires having suitable encoding scheme for the video to allow stream elimination which is implemented by video encoding component as specified in Sect. 7.4. In the following subsections, the components of these building blocks are analysed.

7.3.1 ROMEO Server

The three main building blocks of the ROMEO server are the topology builder, the MTM and the P2P transmitter. In the following sections, we briefly described these modules and their functionalities.

7.3.1.1 Topology Builder

The TB is a software module mainly responsible for creating MMT that effectively construct the P2P overlay network used for the distribution of the multiple streams associated with the visualisation of 3D content. In particular TB performs the following functions at the main server/super-peer. It listens for new peer connection requests. Also, it acts as an authentication proxy (authenticator) for user authentication with the main server. Moreover, it collects network monitoring data from each of the peers. Additionally, it creates multiple P2P application-level multicast trees for content distribution. Finally, TB computes peer insertion on the P2P application-level multicast trees.

When a peer is redirected to a super-peer, it is the responsibility of the TB to compute the peer position at the P2P multicast tree at the access network level.

Initialising the computation process, the TB groups peers according to the requested content. Next, it groups peers according to their proximity to edge routers (ER) geographical aggregation. Finally, it sorts the peers by evaluation.

After grouping and sorting operations, the multiple P2P multicast trees are computed, one for each requested content and ER. Traffic at the ISP core will be transmitted using IP multicast and it will be the ER's responsibility to map each requested content multicast address to specific parent(s) IP addresses(s)—the ER will effectively act as a replicator. To optimise the ER resources, the super-peer predetermines how many top level peers (parents) can be directly fed by one ER. This means that, when constructing each P2P tree, the super-peer arranges a predetermined number of highest ranking peers at the top of the tree and delegates in these the distribution of content to other peers in the same access network. Every time a peer is selected to forward content, its resources are diminished, and if its evaluation becomes lower than other peers, the new highest ranking peer will take the role of parent for subsequent content streaming.

To minimise the issues associated with peers joining and leaving the system, also known as churn, the TB uses the grounding, graceful leaving and redundancy mechanisms. The grounding technique inserts new joining peers always at the bottom of the P2P tree. The algorithm is executed periodically to maximise the efficiency of the P2P tree-promoting and demoting peers. In the graceful leaving concept, the peers, whenever possible, inform the TB about their imminent disconnection, and only stop forwarding content to their children when instructed by the TB or upon a time-out. Lastly, in the redundancy scheme, all peers will be informed of their active parent and a backup parent when inserted on a tree. If the active parent is not reachable within a time-out the peer switches to the backup parent and informs the TB.

The TB specification brings resiliency, scalability and higher performance to ROMEO platform. Resiliency because by using tree separation, a major fault in a specific part of the network will not affect other parts. By grouping peers by common ER, tree depth is significantly reduced, since peers sharing the same access network have improved downstream and upstream bandwidth, which allows more children per parent and thus supporting higher scalability. Finally, improved performance is achieved when the total number of hops between top level parents and their children is significantly lower, which contributes to reduce the packet/chunk delay, jitter and packet loss. Recovery from minor faults (such as peer churn) can also be achieved in a faster way, since the alternative link (backup parent) is on the same access network and ready to start forwarding packets immediately.

7.3.1.2 Multicast Tree Manager

The MTM also runs at the main server/super-peer and it is intrinsically related with the TB operations. MTM collects/aggregates network monitoring data (percentage of packet loss, average delay, jitter and available bandwidth), from all connected peers, provides the TB with peers' updated network conditions. Furthermore, MTM

Client Application

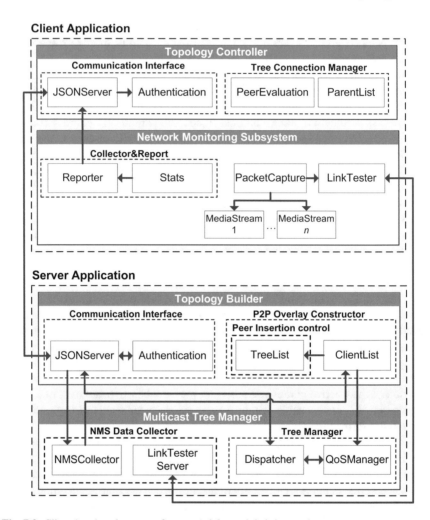

Fig. 7.3 Client (peer) and server software modules and their interactions

allows peers to perform bandwidth tests with the super-peer. Finally, MTM informs the ISP's QoS mechanisms on the endpoints of IP multicast trees at the ISP core network (between the super-peer and the ERs serving the peers).

The MTM performs its functions using the NMSCollector, LinkTesterServer, QoSManager and Dispatcher sub-modules, as depicted in Fig. 7.3. The NMSCollector collects network monitoring data periodically sent by the NMS running at every connected peer. This information is also shared with the TB for quick P2P tree maintenance operations. The LinkTesterServer allows authorised peers to perform bandwidth tests. For the download test, it sends a predefined fixed size binary file, while for the upload test, it expects to receive the exact same file. The results based on this transfer are then used on a composite metric, to compute its evaluation. The

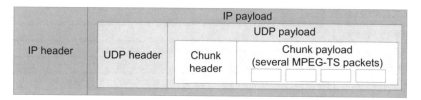

Fig. 7.4 IP packet structure

bandwidth test should be performed at the first time a peer connects (new PeerID) and upon super-peer request (for troubleshooting/maintenance reasons). Moreover, as peers connect to the TB, the QoSManager sub-module is responsible to signal the Internet Resource and Admission Control Subsystem (IRACS), an ISP QoS reservation mechanism on the new peers' addresses, to check if an IP multicast distribution tree exists from the super-peer to the ERs serving the peers (at ISP's core network level). If yes, no changes need to be done at ISP level. If not, either a new IP multicast tree is created or new branches are added to an existing tree. Finally, the dispatcher is used to interface with the TB's JSONServer [13] whenever messages need to be sent or received by the MTM sub-modules.

7.3.1.3 P2P Packetisation/Transmitter

MPEG-2 transport stream (MPEG2-TS) [7] is used for encapsulation since digital video broadcasting—terrestrial second generation (DVB-T2) [6] also uses the same encapsulation method, and the effective use of MPEG2-TS built-in program clock reference makes possible the synchronisation task of the streams received from DVB-T2 and P2P network. As shown in Fig. 7.4, MPEG2-TS packets are then encapsulated as P2P packets to be distributed over IP network. ROMEO employs a packetisation scheme, which aims at efficient and error resilient content delivery over P2P networks. As illustrated in Fig. 7.4, chunk payloads are formed to include MPEG2-TS packets that correspond to a single network abstraction layer unit (NALU)—which are self-contained and decodable units—and chunk headers will carry all the information needed for content aware P2P delivery, chunk selection and retransmission mechanisms. In this way, the P2P packetisation module aims at improving the data delivery performance since other components do not need to extract any specific information within the chunk throughout the streaming over P2P networks.

With the help of the unique chunk identifiers per stream, it is possible to detect missing chunks and request retransmission from the sender with the proposed intelligent chunk selection mechanism. Formation of the small-sized (< MTU size) chunks also improve the performance of the chunk recovery process. Because with the employed scheme, it is possible to request a small-sized single NALU with its unique identifier. Prepared P2P chunks are then delivered through the P2P network.

Table 7.1 P2P header fields

Header parameter	Purpose
Content identifier	Unique id for the content to match with DVB stream
Chunk identifier	Unique id, used for reporting defect P2P chunks
PID (stream id)	Packet ID, used for identifying different elementary streams
PCR	Program clock reference, used for synchronising P2P and DVB-T2 streams
Meta-data flag	If set, the payload contains meta-data (overrides audio flag)
View identifier	Identifies different video views
Descriptor number	Indicates the different descriptors
Layer flag	Indicates the layer type: either base or enhancement layer
Audio flag	If set, the payload contains audio data
Payload size	Size of the payload (number of bytes)
Signature	Digital signature of the payload
Priority	Indicates priority of the stream to decide if the stream is discardable

The P2P transmitter is the component, which delivers the IP streams to ROMEO users over P2P network. It is a multi-threaded UDP sender application that delivers each stream, composed of P2P chunks, in a separate thread that runs in parallel. This module is responsible for encapsulation and decapsulation of the media to be carried over the P2P network. While encapsulating, the data are separated into different streams to be distributed over different multicast trees. In two-layered coding using scalable video coding (SVC) [11], each frame has a base layer bit-stream and an enhancement layer bit-stream. It is possible to separate the original bit-stream and generate two bit-streams, one for base and one for enhancement layer. Besides, audio data and depth information are considered as different streams as well. In order to perform data distribution and synchronisation tasks, some mandatory data is needed for peers and these data, which is listed in Table 7.1 is embedded into the P2P header. As shown in Fig. 7.5, the modules that corresponds to packetisation tasks are part of the main server whereas the modules that depacketise the streams are part of the peer software.

7.3.2 ROMEO Peer

Within the ROMEO peer, the component responsible for topology related actions sends the request to join the ROMEO system with the terminal and network capability information to the ROMEO server. The P2P delivered content is received by P2P Receiver and main stereoscopic content is received by DVB Receiver. Both contents are fed into the synchronisation component for multiple view audio/video synchronisation. The synchronised content is then sent to the audio and video decoders to be decoded and then to the renderers to be rendered at the user display. As a content and network aware system, ROMEO has chunk selection and network

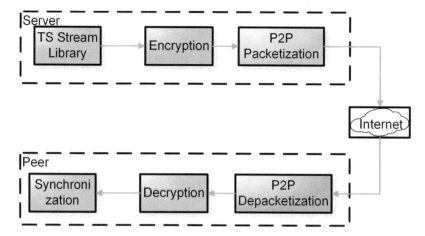

Fig. 7.5 Block diagram for packetisation

monitoring capabilities to monitor the dynamic changes in the network, and adapt to the changing conditions. Also a ROMEO peer can join the collaborating group with A/V communication overlay component and during the collaboration, can share its own content with the other collaborating users with the UGC component.

7.3.2.1 Topology Controller

The topology controller (TC) is a software module that runs at the peer and it is responsible for the P2P functionalities at the peer side. In particular, TC establishes the initial contact with the main server for user authentication and redirection to the nearest super-peer. Furthermore, TC computes the peer evaluation using peer's hardware characteristics and network statistics, as provided by the NMS. Within the responsibilities of TC is to inform the TB of its intention to consume specific 3D content and performing P2P tree operations as commanded by the TB (parent, parent/child or child). Finally, TC establishes connections with parents for content request and accepting connections from children peers for content forwarding.

In order to implement these functionalities the TC uses four sub-modules, as depicted in Fig. 7.3. The authentication sub-module interacts with the user interface and control component for user authentication interactions with the authentication, registration and security component running at the main server. The JSONServer module [13] is responsible for sending and receiving messages to/from the client application modules, whereas the PeerEvaluation module computes the peer evaluation. Lastly, the ParentList sub-module contains this peer's list of parents and backup parents (for each content type) as indicated by the TB.

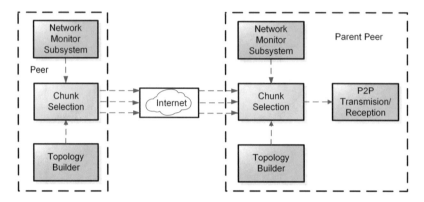

Fig. 7.6 Chunk selection interfaces

7.3.2.2 P2P Reception/ Depacketisation

In each ROMEO peer the P2P Receiver is the module responsible for receiving the P2P streams from its parents. It is a multi-threaded UDP application, which receives each stream by a separate thread in parallel. For each received stream (from P2P network), the P2P Receiver sends the chunks to two different modules: the chunk selection (to selectively forward the chunks to the children peers, as explained in next subsection) and the P2P Depacketisation. The P2P Depacketisation extracts the content from the P2P chunks and feeds the synchronisation component.

7.3.2.3 ROMEO Chunk Selection

The CS module is responsible for the control of the flow of chunks from a peer to its children. The role of CS is two-fold, to adapt the streams to the network conditions in conjunction with the upload capacity of the peer, and to implement a fail-over mechanism that increases delivery of chunks during disconnection events. In doing so, it uses information retrieved from the NMS and the TB. The decision is then passed to the P2P Transmission/Reception module which in turn forwards the selected chunks. As shown in Fig. 7.6, there are two interfaces at the input and two interfaces at the output. These interfaces will transfer data from DVB and P2P and deliver audio and video to decoders. Each view's base and enhancement layers will be transmitted to the same decoder at the same time. This will enable decoders to construct 3D image more efficiently.

The CS module consists of four sub-modules. The InterfaceToNMS requests and retrieves the values of delay, jitter, packet loss and the upload capacity of the peer from the NMS. On the other hand, the InterfaceToTC receives topology updates from the topology controller. Then CS updates the list of peer's children accordingly. Additionally, the BitrateMonitor sub-module measures the bit rate of all the streams that are being received by the peer continuously. The values are used to determine

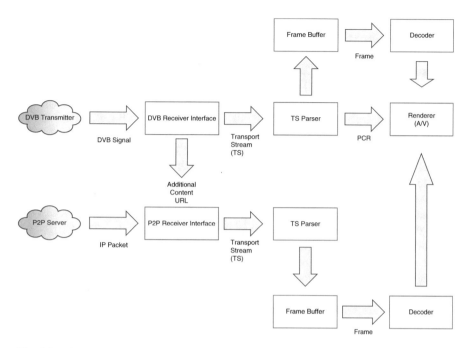

Fig. 7.7 ROMEO synchronisation module block diagram

which streams are going to be forwarded to the children. Last but not least, the ChunkScheduler is the core module of CS, as it determines which streams are going to be forwarded to the children in each tree. This decision-making process takes into account the properties of the stream (e.g. importance, bit rate), the peer capability (e.g. upload capacity), the network conditions (e.g. delay, jitter, packet loss) and the children of the peer in each tree.

7.3.2.4 ROMEO Synchronisation Module

In ROMEO, synchronisation is required at three levels. At a first level, the network synchronisation (play-out synchronisation) takes place, scheduling the play-out time of audio-visual data in order to reduce the jitter and packet loss incurred as packets traverse the network. As a second level synchronisation, the audio and video are synchronised at the terminal (temporal and spatial alignment of 3D video and spatial/object audio) in order to maintain their original temporal relationship. Finally, a third level synchronisation is focusing on the UGC of the A/V communication overlay, where the media playback of different users is synchronised, and the latency differences between users are catered for. Figure 7.7 illustrates the synchronisation module's blocks residing at the receiver.

In this section, we focus on the first and third case. In the case of network synchronisation or play-out synchronisation, the ROMEO peer receives one stream from DVB network and the other streams from the P2P delivery network. Even though both streams have identical presentation time-stamps (PTS), due to different transmission paths, they will not be received at the same time by the peer. The DVB stream packets will, under normal circumstances, be received before the delivery of corresponding stream packets through P2P. To synchronise DVB and P2P streams, a buffer is required. The basic unit extracted from the streams and kept in the buffers is a frame. In ROMEO, each frame has a unique time-stamp within the stream. In addition, in each stream there are also frames with identical time-stamps. In the peer terminal, there is one buffer for each stream. The number of buffers corresponds to the number of streams received by the peer. After all buffers receive their first frames, their time-stamps values are compared. The maximum of the compared time-stamps determines the stream with the maximum lag/delay. The buffers are synchronised with respect to the stream with the highest delay. In other words, frames with earlier time-stamps are dropped in peers that have lower delivery delay.

On the other hand, a clock reference is delivered within the transport stream of DVB transmission, which is used by the audio and video renderers of the 3D media player. Buffering of the frames for synchronisation causes a delay between time-stamps and the clock reference, and the time of some frames may elapse while they are stored in the buffer. To prevent such losses, the synchronisation block adds an offset to the time-stamps of frames at the time of forwarding to the renderers. The amount of this offset depends on the time difference between the arrival time and decoding time of a frame. After these operations, the play-out synchronisation is achieved and a basis for the temporal alignment of 3D video and spatial/object audio is formed.

In the case of user synchronisation, the collaborative experience, which was one of the main goals of ROMEO, raises an additional synchronisation challenge. Due to the nature of P2P transmission, each peer in a group may receive the same frame at different time instances, which causes a delay between peers. To provide a rich and seamless collaborative experience to a group of users, each user must watch the same frame at the same time, or with a negligible (unperceivable) delay difference. ROMEO offers a user synchronisation method similar to the deployed network synchronisation method. This method is implemented before the start of decoding and rendering. After synchronising the received streams for each individual user, the obtained maximum time-stamp values on each peer are sent to the collaboration (A/V overlay) server. The A/V overlay server computes the collaborative delay amongst all collaborating users, and then notifies/synchronises all the peers with the collaborative delay. This delay is computed according to the collaborator with the highest acceptable delay, and sent to each peer in the group.

However, network conditions of several users may differ, which may result in longer delays than practical to compensate. The value of the maximum compensable delay can vary between 5 and 10 s. More than that amount of delay may disturb other peers who have to wait the lagging ones in the group. Therefore, admission

Table 7.2 Updated structure for the NMS data report

Field	Description	Size (in Bytes)
ID	Message ID (identifies this is a report)	4
PeerID	The unique peer ID as given by the TB	32
LocalIP	The local IP address of the Peer (IPv4/v6)	16
NetMask	The local IP subnet mask	4
DL	The download capability in (Kbps)	4
UL	The upload capability (Kbps)	4
nChildren	The total number of children of this peer	4
nCPUcores	The number of CPU cores and its type	4
TotalMem	The size of RAM memory in the peer (MB)	4
FreeMem	The size of available memory (MB)	4
ConsMem	The size of consumed memory (MB)	4
OS	The operating system identification	variable
Delay	The average packet delay (μs)	4
Jitter	The delay jitter (μs)	4
PacketLoss	The packet loss (in %, content specific)	4

control is performed and the users with poor network conditions are not included or accepted in the collaboration group.

7.3.2.5   Network Monitor Subsystem

The NMS has the three important functionalities. It collects peer hardware and software characteristics, in order to determine the end device capabilities. Likewise, NMS collects network traffic statistics (i.e. packet loss, average delay, jitter and available bandwidth) for each received stream. Periodically, it reports the collected data to its parents (it chooses a different parent in each iteration using a round-robin approach) or to the MTM (in case this is a top level peer). In the end, NMS reports can also be triggered by a request from the MTM or when changes in the peer's network conditions cross a specific threshold, for an event-driven approach.

Table 7.2 shows the structure for the periodic NMS report. To save resources and simplify socket management at the receivers, reports are sent to one of the parents, who then collects all the received reports during a time-window and sends all collected reports to its own parent (one by one in a persistent TCP connection). This process goes on until the data gets to the highest peers in the P2P tree hierarchy, who then send the bundle of all collected reports to the NMSCollector sub-module of MTM.

The NMS is also responsible for detecting, reporting and possibly solving peer connectivity problems. If a major failure occurs in stream reception, the NMS of the peer first contacts the NMS of the parent responsible for streaming that specific content, and if reachable, it is up to that parent to solve the problem. If the parent is not able to solve the problem in a pre-specified time-window, or cannot be contacted, the TC is notified in order to immediately switch to the backup parent and inform the MTM.

NMS functionalities are implemented through the four sub-modules, as depicted in Fig. 7.3. Initially, the PacketCapture sub-module passively collects network data such as connection history (start time, duration) or traffic statistics (packet loss, delay and jitter) by exploiting the libpcap library [12]. Moreover, the LinkTester sub-module provides link testing functions with the MTM to determine link characteristics such as the download and upload link capacity. The Stats sub-module is responsible for collecting hardware and software data as depicted in Table 7.2 and for computing the statistics associated with the network data collected by the PacketCapture sub-module. Finally, the so-called Reporter sub-module formats the information collected by the Stats sub-module according to a report template. It is also responsible for sending to the TC's JSONServer [13] the report to the selected peer parent. If this peer is a parent, this sub-module is also responsible for collecting NMS reports from all of its children and to send the report bundle to the selected parent in the hierarchy.

7.4   ROMEO Content Description

In the following section we will describe the 3D video generation process as well as the content packetisation process that enables synchronisation between the broadcast and the broadband access network. The scalable multi-view video encoding module that is placed in the multimedia content server is responsible for compressing the captured and pre-processed raw multi-view plus depth map videos (delivered in YUV 4:2:0 uncompressed format) and subsequently generating a number of bit-streams for each camera view and its sub-layers (e.g. video quality layers and depth). The raw video is provided by the content generation module. Well-known SVC standard [4, 11] is used to individually encode each camera view and its corresponding depth map sequence. The output of this module is sent to the media encapsulation block, where the elementary bit-streams are packed into MPEG2-TS, ready to be encrypted and encapsulated into P2P chunks.

ROMEO deploys a MMT P2P media delivery in a content aware fashion [3]. For increased delivery robustness and for exploiting the multi-path delivery platform, multiple-description extensions are taken into consideration. In order to conserve the underlying video decoding architecture, we employ multiple-description generation such that each description is self-decodable using the standard SVC decoder (each description can comprise multiple quality layers) and a full representation can be obtained easily by post-decoder processing of individual descriptions. This is illustrated in Fig. 7.8. In this hierarchical way, multiple-description feature can easily be added on/removed from the media encoding/decoding complex.

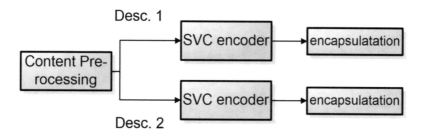

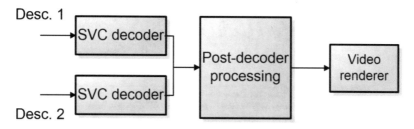

Fig. 7.8 Multiple-description generation

The scalable multi-view video coding approach is inherently scalable in view domain, i.e. any camera view can be decoded independent of any other camera view. In other words, any camera view can be discarded in the network freely, when the network conditions are deteriorating. Another envisaged addition to the scalable multi-view encoding complex is the generation of side information (view interpolation meta-data). The side information is used to assist optimal recovery of lost view frames in the video rendering. The side information is used such that the successfully received camera views' frames are blended in right proportion to obtain the lost/missing views' frames. Hence, the additional data generated in the encoder comprises coefficient sets for blending process and an index sequence to specific coefficient sets at all times. The generation process is independent of the scalable video encoding process, and so is the scalable video decoding process at the peer side. The encoded packets of meta-data blocks will be decompressed in a side process in the core video decoding process, without inter-process communication. The decoded side information (i.e. coefficients and index to them) will be delivered to the multi-view plus depth renderer to use in optimal view interpolation in case of losses.

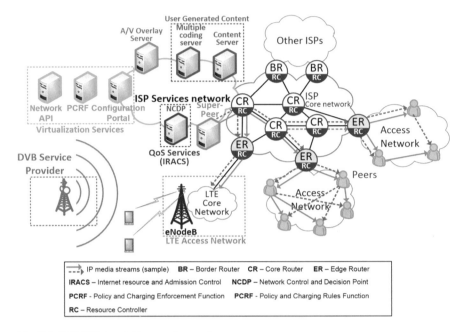

Fig. 7.9 ROMEO data delivery path overview

7.5 ROMEO Data Delivery

This section is dedicated to explain the life cycle of the content from a QoE perspective regarding both the core and the access network including the mobile users. Figure 7.9 provides an overall picture for the networking components architecture, which aims at maximising the experienced quality during the live streaming. ROMEO services are installed in the services network of each ROMEO enabled ISP. Content is then distributed from the ROMEO super-peer to the ROMEO users, passing through the ISP core network and the access network— which can be cabled or wireless. To safeguard user QoE, service requests to use the network infrastructure are subjects to admission control which is performed by the network control decision point (NCDP) and then enforced by the resource controller (RC) module in each ISP core network router. The inter-domain connections are performed based on contracted service level agreements (SLAs) to define the services to be provided and service level specification to include the traffic treatment and performance metrics between the network provider and customers. To ensure these QoS assurances are extended to the user access network, there is an intrinsic relation between the ROMEO super-peer and network API (neutralisation services) who then informs the policy and charging control function (PCRF) [17] to command the ER (also known as the broadband remote access server—BRAS) serving the user to enforce the desired QoS on the access network. The user can also personalise the QoS parameters through the use of the configuration portal. For the mobile users,

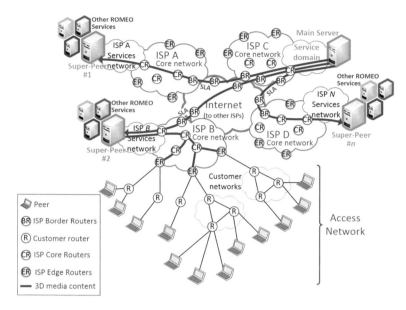

Fig. 7.10 Distribution of ROMEO content at the IP core network

the mobility aware proxy located at the long term evolution (LTE) core network is used to perform the needed adaptations in order to guarantee the best possible user QoE, as will be explained later. Additionally, two other services are provided the A/V overlay server and the UGC component. The A/V overlay server is the responsible to accept and host video conferences between ROMEO users. For each user joining the video conference, the A/V overlay server signals the NCDP to set up the needed QoS assurance at the ISP core network. If users are on different ROMEO enabled ISPs, the QoS is subject to SLAs in place, and the procedures for ISP-to-ISP signalling will be applicable. The UGC server stores and distributes UGC to authorised users. Through the use of the multiple coding server, ROMEO allows administrators to create multiple coding profiles that automatically encode each uploaded content to a specific quality and bit rate.

In an initial stage (small scale deployment) the ROMEO services can be implemented in a single ISP. This facilitates the implementation and dimensioning of the system while providing ISP users the best QoE. Further expansion of ROMEO services can then be achieved by establishing SLAs with additional ISPs. On a larger scale, the ROMEO media streams will be forwarded from the ROMEO main content server to multiple geo-located ROMEO super-peers, as depicted in Fig. 7.10. Synchronisation between the media content database at the ROMEO main server and the media content databases at different super-peers may occur during low traffic conditions or upon request. SLAs between ROMEO enabled ISPs are of utmost importance to guarantee the best QoE to the ROMEO users.

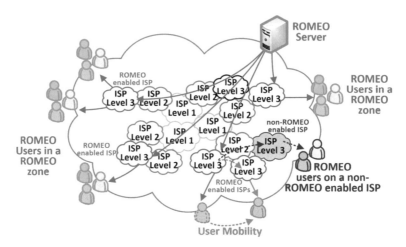

Fig. 7.11 ROMEO user connection scenarios

7.5.1 ROMEO Content Delivery in the IP Core Network

In case a ROMEO user connects from a non-enabled ROMEO ISP, as depicted in Fig. 7.11, the super-peer (at each enabled ROMEO Level 3 ISP) from the closest location will be selected to feed that user. Since the targeted ISP is a non-enabled ROMEO ISP, the border router connecting the ROMEO enabled ISP to the next ISP towards the user will perform the multicast-to-unicast conversion. In such a case the QoS guarantees will be the ones established between the involved ISPs and admission control will be performed.

7.5.2 ROMEO Content Delivery in Access Network

In addition to the QoS enforcement procedures provided by ROMEO within the core network, QoS can also be warranted in the access network segment, in case there are agreements in place (expressed by an SLA) between ROMEO and the ISP operating the fixed access network. In these cases, the ROMEO super-peer notifies the ISP that the content is going to be delivered to a user located in its access network, so the ISP can enforce the expected QoS levels on the access network nodes. The proposed architecture is depicted in Fig. 7.12. The actual network architecture in the access network depends on the standardisation body the network operator chooses to comply with. The standards followed in the ROMEO approach are those furthered by the broadband forum (BBF). The advantage of BBF's standards is two-fold: on one side, most network operators managing fixed accesses comply with BBF recommendation; on the other side, the BBF model has been specified to ensure

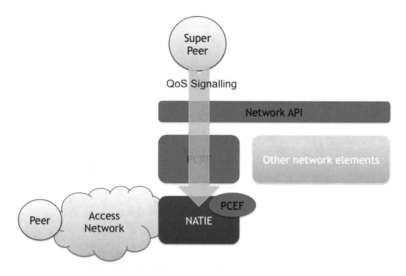

Fig. 7.12 ROMEO access network QoS enforcement

the convergence with the 3GPP standards, which are commonly followed by mobile operators.

To achieve a dynamic QoS assignation, BBF and 3GPP define a policy manager named PCRF [17]. In the ROMEO architecture, a PCRF receives QoS requests from ROMEO upper layer services through its northbound interface (Rx) and enforces them in the NATIE (IP edge router) through its southbound interface (Gx). Since the PCRF is a standardised network element, no new elements are to be introduced in the existing operator's network infrastructure, thus diminishing the impact of ROMEO deployment and therefore facilitating it. The PCRF Rx interface relies on Diameter, which is a complex protocol based on peer sessions. In order to abstract ROMEO from Diameter particularities and simplify the QoS enforcement procedures, an HTML-based overlay has been implemented over the Rx interface. Hence, the ROMEO modules can just request the enforcement of the expected QoS, based on a request/replay sessionless scheme.

7.5.3 Content Delivery for Mobile Users

The solution proposed by ROMEO combines the DVB-T2 broadcast network technology along with access network technologies using a QoE-aware P2P distribution system that operates over wired and wireless links. Inherently P2P networks provide services and functionalities with limited or no centralised infrastructure, as opposed to centralised systems like the IP multimedia subsystem [19]. Hence, ROMEO architecture overcomes the disadvantages of centralised systems, such as lack of redundancy and scalability and single point of failure. Moreover, ROMEO supports full IP mobility to mobile users through the EPC [16], which is the core network of

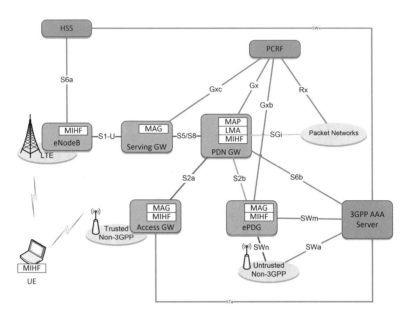

Fig. 7.13 ROMEO mobility modules into LTE/EPC architecture

LTE [9] technology. The protocol that provides IP connectivity using non-3GPP accesses is the proxy mobile IP (PMIP) [10] as described in [1]. The involved entities in the PMIP domain are the local mobility anchor (LMA) and the mobile access gateway (MAG). LMA is the home agent for the mobile node and the topological anchor point for the mobile node's home network IP address. MAG is a function on an access router that manages the mobility-related signalling for a mobile node that is attached to its access link. It is responsible for tracking the mobile node's movements to and from the access link and for signalling the mobile node's LMA. The ROMEO proposed solution architecture is given in Fig. 7.13.

According to 3GPP technical specification 23.402 [1], the LMA functionality is co-located with the packet data network gateway (PDN-GW). The MAG functionalities are implemented at the last hop entities based on the access technology the user equipment (UE) is attached to. In particular, for LTE access, the MAG is a functionality of the serving gateway (S-GW) for trusted WLANs. As such, MAG is implemented at the access gateway (AGW), while for the non-trusted access networks at the evolved packet data gateway (ePDG). Moreover ROMEO extends PMIP to support multi-homing in a similar manner to [2], which is an Active-Internet draft of the Network-Based Mobility Extensions Working Group of IETF. As a result, ROMEO client is not only mobile with the ability to roam in a heterogeneous environment but also capable of receiving simultaneously from multiple interfaces. In order to provide link-layer intelligence and other related network information, as well as to optimise horizontal and vertical handovers, ROMEO exploits the Media Independent Handover (MIH 802.21) protocol [20]. A set of handover-enabling functions called MIH function (MIHF) is used for that purpose. MIHF detects changes in link-layer properties and implements a set of commands that are relevant to handover and switch between links, triggering the IP mobility process of PMIP. The other component is the media independent information service (MIIS) that provides details on the characteristics and services provided by the MIHF from both the mobile node and the network side. The information enables efficient system access and effective handover decisions. The MIIS is also located at PDN GW as this node lies at the heart of the EPC, connecting all other entities. The coupling of PMIP with MIH results in a robust mobility management that guarantees not only reliable handover management, but also efficient network detection and selection. Finally, ROMEO proposes media aware proxy (MAP) as a transparent user-space module used for low delay filtering of scalable multiple-description video streams. Its functionality extends from the network layer to the application layer providing video stream adaptation by taking into account the clients' wireless network conditions. Therefore based on a proposed machine learning model, MAP is able to either drop or forward packets that carry specific layers of a stream to the receiving users. The need for such a middlebox stems from the fact that in a real environment each client must be able to have a guaranteed QoE. The MAP function is integrated with PDN-GW and communicates internally with the MIIS database, ensuring fast transmission rate adaptation decisions, whenever required. It is noted that as PDN GW already implements packet lawful interception and routing, it can easily adopt the MAP functionalities. Subsequently, MAP inherits the EPC scalability and enhances existing processes to guarantee a minimum QoE level.

7.6 Conclusion

This chapter presented an innovative approach for immersive, social TV services as was implemented by the European-funded project ROMEO. In order to support the demanding requirements in terms of bandwidth and QoE of such services, ROMEO exploited the HbbTV concept. The broadcast DVB network was combined with the P2P broadband internet technology to deliver 3D multi-view content alongside UGC in a timely manner. For the P2P tree construction process, the MMT approach was selected for increased delivery robustness and for exploiting the multi-path delivery platform. The SVC standard was used to individually encode each camera view and its corresponding depth map sequence. For synchronising the streams of DVB and P2P networks, the MPEG-TS content packetisation scheme, and particularly the built-in program clock reference, was employed. Finally, the life cycle of the ROMEO content from a QoE perspective was explained, taking into account specific use cases to cater for content delivery to different ISPs as well as to mobile users.

References

1. Architecture enhancements for non-3GPP accesses (2015) 3GPP TS 23.402 V13.3.0
2. Bernardos C (2015) Proxy Mobile IPv6 Extensions to Support Flow Mobility. https://tools.ietf.org/html/draft-ietf-netext-pmipv6-flowmob-14
3. Birkos K, Politis I, Lykourgiotis A, et al (2012) Towards 3D video delivery over heterogeneous networks: the ROMEO approach. In: International conference on telecommunications and multimedia (TEMU), pp 60–65. IEEE, Chania
4. Birkos K, Tselios C, Dagiuklas T, Kotsopoulos S (2013) Peer selection and scheduling of h. 264 svc video over wireless networks. In: 2013 IEEE wireless communications and networking conference (WCNC), pp 1633–1638. IEEE, Shanghai
5. Cesar P, Geerts D (2011) Past, present, and future of social TV: a categorization. In: Consumer communications and networking conference (CCNC), pp 347–351. IEEE, Las Vegas
6. Digital Video Broadcasting (DVB) (2012) Implementation guidelines for a second generation digital terrestrial television broadcasting system (DVB-T2) . TS 102 831 - V1.2.1
7. Digital Video Broadcasting (DVB) (2014) Transport of MPEG-2 TS Based DVB Services over IP Based Networks. ETSI TS 102 034 V1.5.1
8. Digital Video Broadcasting Project (2015) https://www.dvb.org/. Accessed 30 Nov 2015
9. Evolved Universal Terrestrial Radio Access (E-UTRA) and Evolved Universal Terrestrial Radio Access Network (E-UTRAN) (2015) Overall description; Stage 2. 3GPP TS 36.300 V13.1.0
10. Gundavelli S, et al (2008) Proxy Mobile IPv6. IETF RFC 5213
11. Information technology (H. 264/SVC) – Coding of Audio-Visual Objects – Part 10: Advanced Video Coding (2014) ISO/IEC 14496–10, 8th Edition
12. Jacobson V, McCanne S (2009) libpcap: packet capture library. Lawrence Berkeley Laboratory, Berkeley
13. libjson-rpc-cpp package (2015) https://launchpad.net/ubuntu/+source/libjson-rpc-cpp. Accessed 30 Nov 2015
14. Lykourgiotis A, Birkos K, Dagiuklas T, et al (2014) Hybrid broadcast and broadband networks convergence for immersive TV applications. IEEE Wirel Commun Mag 21(3):62–69
15. Montpetit MJ, Klym N, Mirlacher T (2011) The future of IPTV. Multimed Tools Appl 53(3):519–532
16. Network architecture (2015) 3GPP TS 23.002 V13.3.0
17. Policy and charging control architecture (PCRF) (2015) 3GPP TS 23.203 V13.5.1
18. ROMEO (2015) Remote Collaborative Real-Time Multimedia Experience over the Future Internet. http://www.ict-romeo.eu/. Accessed 30 Nov 2015
19. Service requirements for the Internet Protocol (IP) multimedia core network subsystem (IMS) (2015) Stage 1. 3GPP TS 22.228 V13.3.0
20. Standard I, Networks MA (2006) Media Independent Handover Services

Chapter 8
HTTP Adaptive Multiview Video Streaming

Cagri Ozcinar, Erhan Ekmekcioglu, and Ahmet Kondoz

Abstract The advances in video coding and networking technologies have paved the way for the delivery of multiview video (MVV) using HTTP. However, large amounts of data and dynamic network conditions result in frequent network congestion, which may prevent streaming packets from being delivered to the client. As a consequence, the 3D visual experience may well be degraded unless content-aware precautionary mechanisms and adaptation methods are deployed. In this chapter, state-of-the-art HTTP streaming technology and the proposed adaptive MVV streaming method are described. When the client experiences streaming difficulties, it is necessary to perform adaptation. In the proposed streaming scenario, the ideal set of views that are pre-determined by the server according to the overall MVV reconstruction quality constraint is truncated from the delivered MVV stream. The discarded views are then reconstructed using the low-overhead additional metadata, which is calculated by the server and delivered to the receiver. The proposed adaptive 3D MVV streaming scheme is tested using the MPEG dynamic adaptive streaming over HTTP (MPEG-DASH) standard. Tests using the proposed adaptive technique have revealed that the utilisation of the additional metadata in the view reconstruction process significantly improves the perceptual 3D video quality under adverse network conditions.

8.1 Introduction

Recent technological advances in video coding and networked delivery services have made 3D video feasible to be delivered to the clients. 3D video technology has brought a new level of immersion and realism capabilities, taking the user experience beyond the traditional 2D horizon.

C. Ozcinar (✉)
LTCI, CNRS, Télécom ParisTech, Université Paris-Saclay, Paris, France
e-mail: cagri.ozcinar@telecom-paristech.fr; cagriozcinar@gmail.com

E. Ekmekcioglu • A. Kondoz
Institute for Digital Technologies, Loughborough University London, London, UK
e-mail: e.ekmekcioglu@lboro.ac.uk; a.kondoz@lboro.ac.uk

© Springer Science+Business Media New York 2017
A. Kondoz, T. Dagiuklas (eds.), *Connected Media in the Future Internet Era*,
DOI 10.1007/978-1-4939-4026-4_8

The stereoscopic 3D (S3D) video, which is the primary application of 3D video, has achieved extraordinary success in creating a sense of reality. In the S3D technology, two slightly different views are presented to the eyes, which are then merged by the human visual system (HVS) [29] to perceive the 3D effect. However, the S3D technology provides end-users with the illusion of 3D only thought a single view of the scene. Without the possibility of changing viewpoints, S3D technology remains far from being a truly immersive viewing experience.

The 3D experience is further enhanced using multiview video (MVV), which includes more than two views, on multiview displays (e.g. multiview [3], and super multiview (SMV)), where motion parallax [36] are realised. This technology allows end-users to view a scene by freely changing viewpoints, as they do in the real world without wearing special glasses. MVV technology, unlike S3D video, overcomes single view limitation by providing end-users with a realistic 3D viewing experience. However, the acquisition of MVV content is performed with a large number of views. Consequently, efficient video coding/compression and effective streaming techniques are vital.

ISO/IEC MPEG[1] and ITU-T Video Coding Expert Group (VCEG) have been developing video coding tools to standardise efficient 3D video coding solutions. The previous multiview video coding (MVC) [49] and multiview video extension of HEVC (MV-HEVC) standards [52], which focus on the compression of only colour texture views, have not facilitated the generation of additional colour views using depth maps.

The advent of new depth image sensor devices and robust depth estimation tools [43], the multiview-plus-depth (MVD) [27], has widely become trending representation in 3D video research. Depth maps, which are grayscale images—each pixel of a view provides information about the distance of the observed point to the camera—are used to generate additional views.

The new 3D video coding standard, 3D extension of HEVC (3D-HEVC) [28], which was finalised in February 2015, addresses the use of the generation of additional views from some colour texture views associated with their depth maps. The aim is to use multiview display applications with tens of output views from a couple of input camera feeds. However, additional view generation techniques at the receiver side struggle with view synthesised artefacts and occlusion problems.

Regardless of the advances in video coding, delivery of a high volume of sequences is a significantly challenging task, where all captured views—colour texture and depth maps—are required to allow high-quality and realistic 3D scene rendering at the receiver side. MVD streaming with non-adaptive transmission techniques may cause massive congestion in the network, lead to network collapse [15].

---

[1]MPEG is an abbreviation for Moving Picture Experts Group, which is an International Organisation for Standardisation (ISO)/International Electrotechnical Commission (IEC).

The streaming of extremely high volume content over the Internet faces several challenges due to the instantaneously changing bandwidth conditions [24]. Hence, reliable distribution strategies are needed to provide a high-quality MVV streaming experience. In order to tackle this challenge, the goal of this chapter is to deliver MVV to clients/users in a cost-effective manner. To this end, HTTP[2] streaming literature is first reviewed in Sect. 8.2. Next, the proposed adaptive MVV streaming system overview, which is used with the MPEG dynamic adaptive HTTP streaming over HTTP (MPEG-DASH) standard, is described in Sect. 8.3. Section 8.4 presents the experiment evaluations and conclusions are drawn in Sect. 8.5.

8.2 Literature Review

The delivery of video over the Internet has been essential for multimedia communication applications. Two delivery techniques are dominant in the literature: (1) the progressive delivery, (2) the video streaming method, which is more suitable for modern media applications. The video streaming is further classified into two classes, namely the live and on-demand streaming. In live streaming, the video is generated by the server (i.e. transmitter) and simultaneously transmitted to end-users. In doing so, all end-users are watching the same part of the video at the same time as it happens. In on-demand streaming, on the other hand, end-users select and watch to video stream when they want to, rather than having to watch at a particular broadcast time.

In the progressive delivery, the video starts to play when enough data gets downloaded to the viewer's hard drive; this amount is called as a buffer. Although progressive delivery can be delivered to the same protocol as video streaming methods, is not as efficient. It requires the end-user to buffer the high amount of video packet to achieve smooth playback. Most of the downloaded data may not be of use if the end-user does not want to watch the video from the start. Typically it requires an extended period that achieves poor performance for delay-sensitive media services.

In the streaming delivery, the video content consumes less bandwidth than progressive delivery, because only the section of the video that is watched is delivered. Also, it is possible to seek to any point in the video that has not been downloaded. Particularly, HTTP streaming is rapidly becoming one of the most widely used streaming mechanisms to deliver multimedia content over the Internet such as YouTube and Netflix.

In the last decades, HTTP-based media streaming has been widely developed by streaming service providers (e.g. Adobe [1], Microsoft [23] and Apple [2]) to provide high-quality and bitrate adaptive viewing experience to clients. Although

---

[2]Abbreviation of HyperText Transfer Protocol.

these developed services use different terminology, the principle of streaming operations is the same. In the following sub-sections, the state-of-the-art literature review on HTTP video streaming is described in Sect. 8.2.1. Adaptive HTTP streaming technology is presented in Sect. 8.2.2, and MPEG-DASH standard, which is an enabler for efficient and high-quality multimedia delivery over the Internet, is described in Sect. 8.2.3.

8.2.1 HTTP Video Streaming

HTTP streaming, which is a convenience for both the service provider and the end-user, uses the TCP and the Internet Protocol (IP). TCP is a connection-oriented protocol that offers reliable transmission. UDP, on the other hand, is a connectionless protocol, which means there is no mechanism to set up a link between the service provider and end-user. Therefore, UDP does not provide a guarantee that streaming packets sent would reach the end-user. Firewalls or proxies may also block UDP packets from streaming due to unfamiliar port numbers. TCP packets, however, may lead unacceptable increased transmission delay in the packet-loss sensitive environments.

A dominant share of current streaming traffic is being delivered using HTTP/TCP. HTTP streaming has several benefits for both streaming provider and the end-user. First, it allows for reuse of the existing Internet infrastructure; consequently, it offers better scalability and cost effectiveness for the service providers. It is firewall friendly because most firewalls are configured to support its outgoing port number, i.e. TCP port 88. HTTP streaming prevents packet losses using TCP's packet re-transmission property. HTTP user can also access the HTTP web server directly, and it may use the advantage of HTTP caching mechanism on the Internet to improve its streaming performance. Therefore, it does not require special proxies or caches.

Early HTTP streaming mechanisms were based on progressive delivery (*see* Sect. 8.2) that only allows the end-user to access the video before the data transmission is completed. In order to support streaming features (e.g. rewind, fast forward, pause, etc.), the trick mode was adopted in the HTTP streaming mechanism that diverts downloading to the video segment file that the end-user has requested. To do this, the video file needs to be segmented that is a significant development for the current standard adaptive streaming concept.

Video segmentation, allows the end-user to request to different video segmented content, enables reducing the overall transmission delay [37] and adapts the video quality based on the changing network conditions to ensure the best possible viewer experience [30, 35].

8.2.2 Adaptive HTTP Streaming

Most of the current networked video services run on a best-effort delivery basis, meaning that there are no guarantees of specific performance, but the network protocol makes a best effort to deliver the transmission packets to their destination. The main need for these networked services is the ability to manage the video content to the network (i.e. the Internet).

In order to address progressive delivery limitations, adaptive bitrate mechanism was developed by streaming providers to use the available network resources efficiently and provide high-quality video viewing experience. Adaptive video streaming system also guarantees the highest video quality possible without a need for long buffering [35].

Adaptive streaming, which is an essential link between media content and the end-user to yield enhanced video experience for end-users, adjusts the media transmission bandwidth based on the changing network conditions. Most adaptive streaming techniques provide the optimum video playback quality [34] and minimises start-up delay [35]. The media contents are available in multiple encoded segments at the server before the transmitting. The end-user regularly measures the available bandwidth [26] and/or buffer occupancy level [30] and request the next segmented video in an appropriate bitrate.

In the last decade, several adaptive streaming mechanisms have been developed such as Adobe's HTTP Dynamic Streaming [1], Apple's HTTP Live Streaming [2] and Microsoft's Smooth Streaming [23]. Additionally, a large portion of video content is delivered through HTTP server-based streaming services. This increasing momentum has resulted in the standardisation of MPEG-DASH [48] in April 2012.

8.2.3 MPEG Dynamic Adaptive HTTP Streaming

To solve the problems in the Internet video streaming, HTTP-based adaptive bitrate streaming solution, which is MPEG-DASH [48], is standardised under ISO/IEC MPEG [13]. The MPEG-DASH specifications are contained in four following parts: (1) media presentation and segment formats, (2) conformance and reference software, (3) implementation guidelines and (4) segment encryption and authentication.

The MPEG-DASH standard, where the complete streaming logic is at the receiver, provides an enhanced QoE enabled by intelligence adaptation to different link conditions, device capabilities and content characteristics. According to the bandwidth measurement or the client's hardware specifications, the most appropriate bitrate [55] or video quality [6] is selected to be played. The MPEG-DASH standard can use media segments that are encoded at different resolutions, bitrates and frame-rates. The DASH client can individually request these multiple self-decodable media streams through HTTP. Therefore, it enables the end-user to switch between different qualities, resolutions and frame-rate. To play the media, the DASH client requests the segments one after the other one, when only needing them.

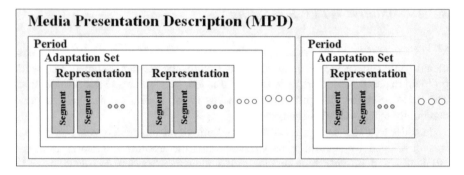

Fig. 8.1 The high-level data model of the media presentation description (MPD)

System description of the MPEG-DASH standard specifies extensible markup language (XML) and binary formats that enable media streaming from HTTP servers to HTTP clients. The standard primarily defines two formats: (1) the media presentation description (MPD), which is a DASH media presentation that can be downloaded by the HTTP client. (2) The segment formats that can be requested with the help of HTTP GET and HTTP partially GET using HTTP/1.1 as defined in RFC 2616 [8].

The high-level data model of the MPD is shown in Fig. 8.1. The MPD, which is downloaded by the DASH client before playback, contains a description of all the information that the video player need to play video sequences. The MPD describes the sequence of periods in terms of time. Each period itself has multiple representations that consist of a set of different available bitrates, resolutions, codecs, captions, languages, etc. For example, audio components in various languages, different video components providing various views of the same programme, subtitles in multiple languages. Within a period, video descriptions are arranged into adaptation sets, which contain a set of representations of the video. Also, adaptation sets are comprised of representations that contain encoded video segments that can be individually requested by the end-user via a TCP request. Also, in order to access segments, a URL address is provided for each segment.

8.2.4   Future Research Directions

There has been an ongoing research on defining and developing the future of Internet architectures, such as cache and forward networking (CnF) [32], NetInf [5] and content-centric networking (CCN) [16]. CCN, which is an emerging Information-Centric Networking (ICN) project, is used as content identifier rather than host identifier. As a result, video sequences can be stored anywhere in the system that can achieve an efficient content distribution. The project allows the network to caching video by attaching storage to routers. Accordingly, the routers become both forwarding elements and caching elements.

In parallel with technical developments in the Internet architecture, adaptive streaming solutions are emerging for the future Internet. For instance, Liu et al. proposed dynamic adaptive streaming over CCN [53], which shows that MPEG-DASH can be adapted relatively easily to a CCN environment taking advantage of the caching features offered by CCN.

At the time this chapter is written, there is much ongoing work in the framework of the MPEG standardisation activities on MPEG-DASH, which is progressing toward its 3rd edition. The integration of new 3D video formats (namely, AVC-based, MVC-based and HEVC-based) includes depth data in MPEG-DASH [10], spatial relationship description, access control, multiple MPDs, full duplex protocols (i.e. HTTP/2), authentication, generalised URL parameters, advanced and generalised HTTP feedback information activities are progressively ongoing. MPEG-DASH currently conducts five core experiments in its 3rd edition [9]:

1. Server and network assisted DASH (SAND) [47].
2. DASH over full duplex HTTP-based protocols (FDH) [41].
3. URI signing for DASH (CE-USD) [46].
4. SAP-independent segment signalling (SISSI) [45].
5. Content aggregation and playback control (CAPCO) [50].

However, the MPEG-DASH standard does not cover the design of the MVV adaptation strategy. In this chapter, we propose an enhanced adaptive MVV streaming technique and show the advantage of the proposed approach both regarding objective and subjective quality. The proposed adaptation strategy considers the use of MPEG-DASH over HTTP 1.1, with the adaptation logic performed on the end-user side.

8.3 System Overview

Figure 8.2 shows a schematic diagram of the proposed HTTP delivery system. The captured MVD content is divided into equal-length of temporal segments, encoded by the HEVC encoder at various bitrates and stored as self-decodable single-layer streams in the HTTP server. The solution does not impact the core codec algorithm and is codec agnostic. The MVV streams exist in the server in two parts: MPD (*see* Sect. 8.2.3) that contains the manifest of the adaptation strategy (*see* Sect. 8.4.3) and MVD segments that contain the MVD streams in the form of segments. Each segment contains the size of a group of picture (GOP), which is typically around 0.5 s.

The end-user retrieves MVD segments and, as necessary, the content is adapted to the dynamic network conditions by discarding the pre-determined view(s). View discarding order, which is computed by the video encoder, is transmitted with the MPD to the end-user each five-second period. Network adaptation is performed on the receiver side, which relies on the adaptation rule transmitted by the server, and then retrieves the corresponding side information (SI) for MVC reconstruction.

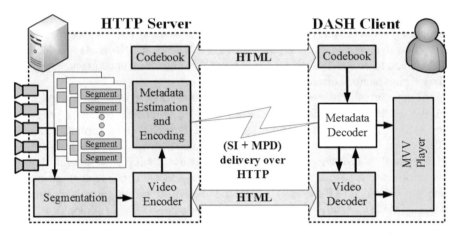

Fig. 8.2 Overview of the proposed delivery system

Discarded views are reconstructed by incorporating the low-overhead SI, which is requested by the user at times of adaptation. The SI stream contains the codebook index values of the delivered views weighting factors, which are calculated using cross-correlation method. The codebook, which is created as a result of weighting factor estimation (i.e. metadata estimation in Sect. 8.4.1) at the encoder, is downloaded by the user during the start-up buffering period.

After downloading the codebook, to play the MVV content, the DASH client should obtain the pre-estimated MPD through HTTP and selects the appropriate subset of encoded views. For each view that can potentially be discarded as a result of the instantaneously available network capacity, the DASH client continuously requests the segments from the HTTP server and monitors the network throughput. Depending on the network throughput and discarding order in the MPD file, the DASH client decides on whether to adapt by requesting a subset of views or a complete set of views.

The information that is needed to reconstruct the discarded views as a result of adaptation is estimated according to the metadata estimation process with the help of the DIBR technique. The estimated metadata stream is then delivered as SI, the details of which are explained in Sect. 8.4.1. The SI is utilised to reconstruct discarded views with high quality at the receiver side, which is delivered to the DASH client using the `HTTP GET` [8] method.

8.4 Adaptive MVV Reconstruction

In this section, the key components of the proposed adaptive MVV reconstruction model for MVV streaming are described as follows: Sect. 8.4.1 presents the metadata (i.e. SI) estimation process. The details of the quadtree-based adaptive block-size selection procedure are explained in Sect. 8.4.2. Finally, Sect. 8.4.3 presents the dynamic view adaptation strategy.

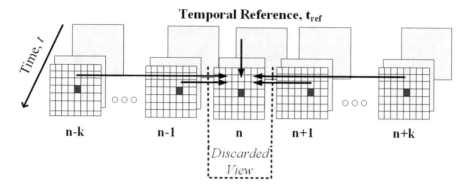

Fig. 8.3 Cross-correlation method for N number of reference views, which is equal to $2k + 1$. In the process of recovering of discarded views, its neighbouring delivered views from both directions are utilised

8.4.1 Metadata Estimation

To estimate the additional metadata for recovering frames within a discarded view, the co-located frame from the delivered views, V_{del}, and a previous temporal reference, t_{ref}, within the same discarded view(s) are utilised, as shown in Fig. 8.3. V_k and V_{k+N} are the most side views of the complete ensemble of MVV content. Hence, $V_{all} = \{ V_k, \ldots, V_{k+N} \}$ where V_{all} is available view domains. The cross-correlation [38] and depth image-based rendering (DIBR) techniques are utilised in the server in such a way that the depth-aided image reconstruction quality is superior to that of the original view. In the transmitter side, for each view assumed to be temporarily discarded from streaming, all corresponding blocks are replenished to calculate weighting factors. The weighted sum of all projected corresponding blocks from delivered views is computed as:

$$\widehat{B}_s(x, y, t) = \sum_{c \in V_{del}} \left[\widetilde{B}_c(x, y, t) \cdot w_c \right] + \sum_{c \notin V_{del}} \left[B_c(x, y, t_{ref}) \cdot w_c \right] \tag{8.1}$$

where $\widehat{B}_s(x, y, t)$ represents the reconstructed pixel at (x, y, t). x and y are the horizontal and vertical coordinates of the pixel. t is the current frame time. V_{del} is delivered view domains. $\widetilde{B}_c(x, y, t)$ represents the DIBR projected block of the cth view, and $B_c(x, y, t_{ref})$ is the temporal reference of the target view block, which is the last frame's block in the corresponding previous temporal segment. t_{ref} is the temporal reference time, and w_c represents the weighting factors of each block for each view, which corresponds to the SI.

This model, as shown in Fig. 8.3, recovers the discarded views, V_s where $s \neq \{k, k+N\}$, with the smallest possible pixel error in relation to its uncompressed original representation. In order to estimate discarded view(s), the sum of squared errors, e_s^2, between the reconstructed pixel values, $\widehat{B}_s(x, y, t)$, and original pixel values, $B_s(x, y, t)$, is calculated as shown in Eq. (8.2):

$$e_s^2 = \sum_{x=1}^{X} \sum_{y=1}^{Y} \left[\widehat{B}_s(x, y, t) - B_s(x, y, t) \right]^2 \tag{8.2}$$

where X and Y represent the width and height of the current block, respectively. In order to minimise e_s^2, it is necessary that the derivative of e_s^2 with respect to each of the weighting factors w is equal to zero, i.e.

$$E\left[2 \cdot e_s \frac{de_s}{dw}\right] = 0 \quad for \quad every \quad w \quad then, \tag{8.3}$$

$$\frac{de_s}{dw_s} = \widetilde{B}_s \quad , \quad s \in V_{\text{del}} \quad thus, \tag{8.4}$$

$$2 \cdot E\left[e_s \frac{de_s}{dw}\right] = 2 \cdot E\left[e_s \widetilde{B}_s\right] = 0 \quad s \in V_{\text{del}} \tag{8.5}$$

Neglecting the constant numbers,

$$E\left[\left(\widetilde{B}_k \cdot w_k + \ldots + \widetilde{B}_{k+N} \cdot w_{k+N}\right) \cdot \widetilde{B}_k\right] = 0$$
$$\vdots \tag{8.6}$$
$$E\left[\left(\widetilde{B}_k \cdot w_k + \ldots + \widetilde{B}_{k+N} \cdot w_{k+N}\right) \cdot \widetilde{B}_{k+N}\right] = 0$$

Hence,

$$\begin{bmatrix} E\left[\widetilde{B}_k \cdot \widetilde{B}_k\right] & \cdots & E\left[\widetilde{B}_k \cdot \widetilde{B}_{k+N}\right] \\ \vdots & \vdots & \vdots \\ E\left[\widetilde{B}_{k+N} \cdot \widetilde{B}_k\right] & \cdots & E\left[\widetilde{B}_{k+N} \cdot \widetilde{B}_{k+N}\right] \end{bmatrix} \cdot \begin{bmatrix} w_k \\ \vdots \\ w_{k+N} \end{bmatrix} = \begin{bmatrix} E\left[\widetilde{B}_k \cdot \widetilde{B}_s\right] \\ \vdots \\ E\left[\widetilde{B}_{k+N} \cdot \widetilde{B}_s\right] \end{bmatrix} \tag{8.7}$$

where $E\left[\cdot\right]$ represents the normalised expected value [19].

After estimating the weighting factors per block for all available views, to design the codebook, the k-means clustering algorithm [18] is applied to the estimated weighing coefficients. Each weighting coefficients (or factors) (w) forms coefficient vectors in the codebook, which is encoded using an l-bit codeword, as described in Eq. (8.8).

$$W_i = \left[w_{N-k} \ldots w_N \ldots w_{N+k}\right] \tag{8.8}$$

where coefficient vector W is chosen from a finite set of coefficient vectors in the codebook with size L, and $L = 2^l$. Also, the index number is denoted as i $(1 \le i \le L)$.

The codebook is downloaded from the HTTP server by the DASH client at the beginning of streaming. The index numbers of each computed coefficient vector corresponding to each computed block are embedded in the SI stream in the HTTP server. The DASH client parses the codebook index values embedded in the SI

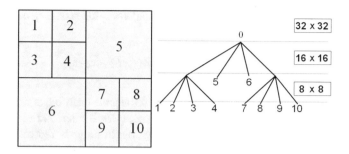

Fig. 8.4 A variable block-size and the corresponding nested quadtree structure

stream to recover the corresponding coefficient vectors from the codebook. For the reconstruct of discarded view(s), correctly received and decoded frames of neighbouring views' with the corresponding temporal reference frame are utilised. The reconstruction is performed as per the weighted summation in Eq. (8.1). This reconstruction is applied for each discarded segment of the view, which is the size of a GOP.

8.4.2   Quadtree-Based Adaptive Block-Size Selection

Quadtree coding [11, 39], which has been widely used for block partitioning in video processing to take advantage of variable block-size, is employed in the proposed view reconstruction mechanism. This technique segments the regions in such a way that they can be reconstructed at high quality with low overhead. The block partitioning process is completely independent from the partition used in the underlying codec.

Figure 8.4 illustrates the exemplary quadtree structure. As can be seen, 8×8 and 32×32 are chosen as the smallest and largest block-sizes, respectively. However, smaller block-sizes, used for regions that can be reconstructed at low quality, increase SI overhead. For this purpose, the trade-off between the discarded views' reconstruction quality and the resulting SI overhead size are optimised using the Lagrangian optimisation method [40]. At that point, optimum block-sizes for each frame of discarded views are calculated.

The block partitioning method evaluates different block-sizes adaptively and assigns an optimum block-size for each region in the frame. Each region is divided into four equal size blocks starting from the largest block-size (i.e. 32×32 block) in a top-down approach [22].

In the block-size optimisation, the overall block distortion, D, is minimised subject to a limited overall SI rate-budget $B_{\max}$. The value of $B_{\max}$ was calculated experimentally through subjective training using three different MVV contents (*Café*, *Newspaper* and *Book Arrival*). In this method, the cost of each possible block-size is calculated by Eq. (8.9), and the smallest value is chosen as the optimal for each block.

$$\underset{x}{\arg\min} \, J(b) = D(s) + \lambda \cdot B(s), B(s) < B_{\max} \tag{8.9}$$

where J is expressed as the cost value, and b represents each block number in the quadtree structure (*see* Fig. 8.4). λ is the Lagrangian multiplier, and $B(s)$ is the cost of transmitting the additional metadata. Every partitioning is represented by the corresponding codevector s, which is assigned a variable length codeword from a given quadtree structure. l denotes the code length; when the cost of transmitting the quadtree structure is included, the overall cost function becomes

$$J(x) = \sum_{x=0}^{P-1} \sum_{y=0}^{P-1} D(x, y) + \lambda_1 \cdot l + \lambda_2 \cdot Q \tag{8.10}$$

where P represents the block-size (such as *32, 16, 8*) in the view, λ_1 and λ_2 correspond to the Lagrangian multiplier values for the SI and quadtree structure overhead, respectively. λ_1 and λ_2 are obtained experimentally through subjective training. Three different MVV contents, *Café*, *News paper* and *Book Arrival*, were utilised to estimate the optimum λ_1 and λ_2. Q denotes as the number of bits (quadtree code-length) required for the signalling of the quadtree structure. l is the number of bits necessary for the SI, which is explained in Sect. 8.4.1.

In this work, the mean square error (MSE) based distortion metric has been utilised in $D(x, y)$. However, other perceptual quality metrics (e.g. Synthesised View Distortion Change (SVDC) [44], Structural SIMilarity (SSIM) [51], Spatial Peak Signal to Perceptual-Noise Ratio (SPSPNR) [54]) are equally compatible with the proposed system.

8.4.3 View Adaptation Using HTTP

In order to cope with the varying network conditions during streaming, two possible adaptation strategies can be employed in the outlined system model. Firstly, it is possible to reduce the streamed views bitrate symmetrically to match the instantaneous changing bandwidth. However, this method would lead to a reduction in the reconstruction quality of all received views and depth maps, thus resulting in a decrease in the perceptual quality. The second option is to transmit selective view streams through content-aware bandwidth adaptation and to allow the missing (discarded) view(s) to be reconstructed at the receiver using the delivered views. This strategy would not end up in compromising the delivery quality of some views, unlike the first strategy.

The proposed system, which bases on the second described strategy, reduces the transmitted number of views during the network congestion period. The proposed system aims at enabling the receiver to reconstruct all required views at the highest

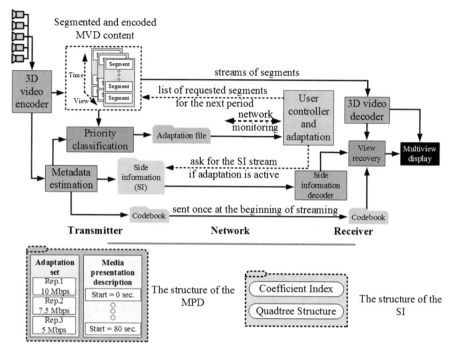

Fig. 8.5 The proposed adaptive MVV delivery scheme. The SI stream is delivered as the overhead, which contains weighting factors (Sect. 8.4.1), and quadtree structures (*see* Sect. 8.4.2)

possible quality at all times by incorporating the low-overhead SI. In contrast to the first explained strategy (i.e. symmetrical quality reduction), this method cannot result in compromising the perceptual quality of all encoded views. However, it leads to maintain the highest possible quality at the receiver side, giving the extraction and usage of optimised SI along with the delivered high-quality views and depth maps.

Figure 8.5 shows the proposed adaptive MVV delivery scheme. All adaptation parameters are prepared according to the possible network bandwidth conditions, i.e. r_{net}, at the server, and they are inserted to the MPD.

MPD, which is prepared at the transmitter, is downloaded by the end-user (receiver) before the streaming starts and updates are delivered when necessary (e.g. content changing significantly). MPD is prepared according to the possible network bandwidth conditions, i.e. r_{net}, at the server. All adaptation parameters are inserted to the MPD. According to the user bandwidth, the user's receiver reads the MPD and request optimum bitstreams.

In order to prepare MPD, the number of views is discarded and evaluated at the transmitter. Consequently, optimal discardable view(s) are determined that minimise the overall MVV distortion to meet network bandwidth (r_{net}). This process can be formulated as an optimisation function, $f(\cdot)$, which is defined as in Eq. (8.11):

$$f(\text{Dist}_k, \ldots, \text{Dist}_{k+N}) \quad and \quad satisfy \quad (R_V + R_{\text{SI}}) < r_{\text{net}} \qquad (8.11)$$

where, Dist_k, $\ldots$, Dist_{k+N} are objective distortion of views. R_{SI} is denoted as the additional metadata overhead for the proposed adaptation and the total encoded MVV content bitrate is R_V. The total bitrate requirement of the proposed adaptation, $R_V + R_{\text{SI}}$, and the overall distortion, $\text{Dist}_k + \ldots + \text{Dist}_{k+N}$, are considered for each segment, which is typically around 0.5 s.

Delivered view(s) are determined using the priority information, which is the process to minimise the overall MVV distortion subject to the r_{net}. The priority estimation is described in Eq. (8.12):

$$p_s = \frac{\omega_s}{\sum_{j=k}^{k+N} \omega_j} \qquad (8.12)$$

where p_s is denoted as the view prioritisation, s belongs to a discrete set of views between (and including) views k and $k+N$. Also, ω_j is expressed as the priority weight of each view.

The priority weight of each view is determined through classification with an aim of minimising the overall reconstruction distortion subject to the bitrate budget (r_{net}). The overall cost minimisation function is described in Eq. (8.13).

$$\underset{k}{\arg\min} P(k) = D_s(k) + \lambda \cdot R(k), \quad R(k) < R_a \qquad (8.13)$$

where $R(k)$ is the overall transmitted bitrate after discarding some of the views (including the bitrate of the SI, R_{SI}), $D_s(k)$ is the average reconstruction distortion of all discarded view(s) estimated using MSE. Also, λ is the Lagrangian multiplier, which is set through subjective experiments using three different MVV contents (*Café*, *Newspaper* and `Book Arrival`).

Each view is encoded using the similar coding parameters (e.g. similar QPs) and stored in the HTTP server. At this point, the user can manage the streaming session based on the MPD, which contains adaptation information. Hence, the pre-determined subset of views can be requested using the *HTTP GET* method when the available network bandwidth cannot accommodate the transmission of the complete collection of encoded MVV. Depending on the measured bandwidth, some views are effectively discarded to be recovered using delivered views with the SI at the receiver. In the most severe condition, to be able to recover all views within the total baseline of the MVV set, the edge views (i.e. side-most views) and their associated depth information need to be delivered.

8.5 Experiment Results

In this section, experiment results are presented to demonstrate the enhancement effects of the proposed adaptation mechanism for MVV streaming. Thus, both objective and subjective results are depicted to highlight the positive impact of the proposed approach over the congested network.

Table 8.1 MVV Test Contents Specifications

Sequence	Camera spacing (cm)	Resolution	Selected views
BookArrival	6.5	1024×768	6,7,8,9,10
Newspaper	5	1024×768	2,3,4,5,6
Café	6.5	1920×1080	1,2,3,4,5

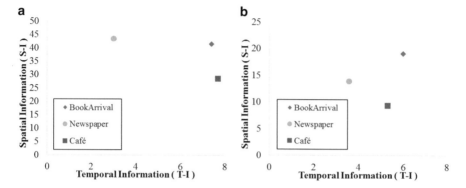

Fig. 8.6 Spatial-Information (S-I) versus Temporal-Information (T-I) indexes of the selected MVV test sequences. (**a**) For texture. (**b**) For depth map

8.5.1 Experiment Setup

The experiments were conducted using five adjacent colour texture views and depth maps (i.e. $N = 5$ where N is the number of available views) from three different MVV test sequences, which are *Book Arrival* [7], *Newspaper* [12] and *Café* [17]. The properties of these contents are summarised in Table 8.1. Figure 8.6 also shows the Spatial-Information (S-I) and Temporal-Information (T-I) indexes on the luminance component of each sequence, as described in the ITU-R P.910 [31].

Figure 8.6a presents the S-I and T-I indexes for texture view. From the figure, it can be inferred that *Book Arrival* and *Newspaper* have large S-I values for texture view, whereas *Café* has a small value, i.e. low spatial details. Also, the *Café* and *Book Arrival* sequences contain higher T-I value for both texture and depth maps, which indicates high-level temporal detail.

HEVC and MVC standards were utilised to analyse the performance of the proposed adaptation system. Therefore, HM v10.1 was used to encode each MVV sequence. Various QPs were selected for each colour texture sequence, which were 20, 26, 32 and 38. Also, the depth maps bitrates were chosen to be equal to the percentage of 20 % of the colour texture bitrates. Hence, appropriate compressed depth maps were chosen by trials. Hierarchical B pictures were used with a GOP length of 16, and a single GOP was inserted into each video segment. In the case of MVC, the inter-view prediction structure was determined based on the view discarding pattern.

To further evaluate the performance of the proposed approach, MPEG DASH [20] software was incorporated in the sever–client setup [26]. Three regular PC have been used in this setup. A PC has been used as a transmitter (i.e. streaming server), one regular PC as a receiver (i.e. streaming client), and one regular PC as a router based on the Dummynet tool [33] to emulate network environment.

To evaluate the performance of the proposed approach, MPEG View Synthesis Reference Software VSRS v3.5 [42] was incorporated as an adaptation reference. In MPEG VSRS, the two nearest left and right adjacent views are utilised in this reference to synthesise discarded view(s). MPEG VSRS was used as the base to estimate the views that are not delivered along with the received SI and various codebook sizes. The additional overhead from the proposed method (i.e. SI) is included in all reported results.

Experiment results are compared using the Bjøntegaard Delta (BD) method [4], which describes the distance between two RD curves. In this manner, PSNR difference, namely ΔP, in dB averaged over the entire range of bitrates, and bitrate difference, namely ΔR, in percentage averaged over the entire range of PSNR, are identified.

Furthermore, subjective tests were performed [21] in accordance with the ITU-R BT.500 [14]. In total, 18 non-expert observers (12 males and six females), aged between 26 and 45, participated in the test. Each test session started after a short training and instruction session. Each subjective assessment session lasted up to half an hour.

The Double Stimulus Continuous Quality Scale (DSCQS) was utilised using a scale that ranged from 0 (Bad) to 100 (Excellent). Observers were provided the freedom to view video pairs (original reference and processed) over and over again as they wish before reaching a verdict. The collected Mean Opinion Scores (MOS) were then analysed for each subject across different test conditions using the chi-square test [25], which verified MOS distribution consistency. Following this process, the outlier was detected and discarded from the reported scores.

8.5.2   Results and Discussion

In this section, the experiment results are given along with a discussion. Also, both objective and subjective results are demonstrated. The results are presented in a comparative manner, which clearly shows the proposed approach performance compared with the reference.

8.5.2.1   Effect of Codebook Size Used on Reconstructed View

The view reconstruction performance of the proposed approach using different codebook sizes is demonstrated using RD curves. Figure 8.7 shows the resulting average view reconstruction performance using different codebook sizes for *Book Arrival*, *News paper* and *Café*, respectively.

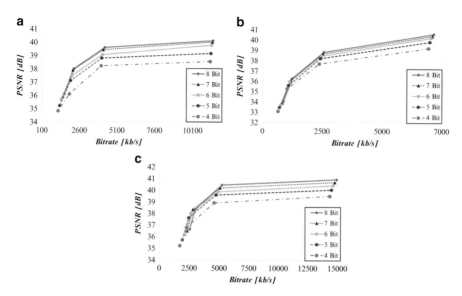

Fig. 8.7 Missing view reconstruction performance using different codebook sizes. (**a**) RD curves for *BookArrival*. (**b**) RD curves for *Newspaper*. (**c**) RD curves for *Café*

In this experiment, only the same view was discarded for each tested codebook at a time, and the average view estimation quality in terms of PSNR was then calculated from the discarded views. The estimated image quality was calculated with respect to the uncompressed original view. The bitrate includes the transmitted MVV content and the overhead caused by transferring the additional SI. Three different codebooks created for each discarded view, which contains a varying number of coefficient vectors (W), are described by the number of bits (l-bit). Each view was assumed to be discarded, and all corresponding blocks were estimated using different size codebooks. The objective reconstruction quality for each view was compared to others in order to determine the optimum views to be discarded for each period.

As can be seen in the results presented in Fig. 8.7, the performance of the proposed approach tends to saturate each of the content pieces when the size of the used codebook increases. However, a relatively large performance gap is obtained between the five-bit and four-bit codebooks. Accordingly, a five-bit codebook achieves view reconstruction performance closer to that of the largest codebook sizes, and benefits from the advantage of lower SI overhead. To this end, five-bit codebook was used in the remaining experiments.

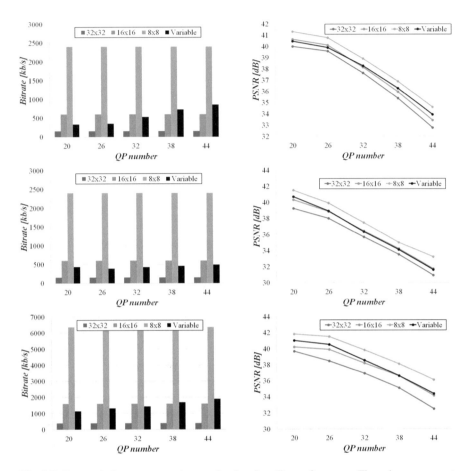

Fig. 8.8 Proposed view reconstruction overhead and quality performance. The colours represent calculated bitrate and quality performance for a block-size of 32×32, 16×16, 8×8, and variable block-size

8.5.2.2 Effect of Block-Size Used on Estimated View

Figure 8.8 illustrates the performance comparison of the proposed reconstruction method using different block-sizes. The aim of this analysis is to demonstrate the effectiveness of the variable block-size selection process (*see* Sect. 8.4.2).

In this experiment, 32×32, 16×16 and 8×8 fixed block-sizes were compared to the quadtree-based variable block-size. As can be seen in the figure, for the *Book Arrival* and *News paper*, (1024×768) sequences, 32×32, 16×16 and 8×8 require 153.6, 614.4 and 2457.6 kbps, respectively. Furthermore, the high-resolution (1920 × 1080) *Café* sequence requires 405, 1620 and 6480 kbps in order to reconstruct a discarding view for 32×32, 16×16 and 8×8, respectively.

As can be seen in Fig. 8.8, the resulting overhead from the quadtree-based variable block-size demonstrates an increasing bitrate performance, whereas video distortion increases. The reason for this is that increasing video distortions present poor reconstruction performance, and thus the proposed approach requires higher amount of metadata (i.e. SI) to construct a missing view with high quality. Moreover, it is observed that corrupted depth maps affect the required metadata overhead size. For instance, the *Café* sequence, which contains inaccurate depth maps compared with others, requires high amount of metadata.

Furthermore, the proposed view reconstruction quality performance in Fig. 8.8b, d, f supports the analysis of view estimation overhead in Fig. 8.8a, c, e. As the block-size increases, the objective quality is also enhanced. However, the increase in the overhead of the view estimation metadata, i.e. SI, reduces the coding performance in the RD curves. In the proposed view reconstruction approach, each block is assigned a weighting coefficient and increased SI overhead decreases the degree of coding efficiency. For this reason, a quadtree-based variable block-size selection mechanism employs cost-quality optimisation in an exchange between SI overhead (cost) and the view reconstruction quality.

Experiment results for three MVV sequences indicate that the quadtree-based variable block-size shows optimum performance compared with the fixed block-sizes (e.g. 32×32, 16×16 and 8×8). This occurs primarily because of the variable block-size selection process that exploits the quadtree structure and reduces SI overhead.

8.5.2.3   View Reconstruction Performance

In order to further investigate and evaluate the performance of the proposed view reconstruction, another view estimation method is incorporated as an additional reference: MPEG VSRS without using SI. In this reference, the nearest left and right neighbouring views are projected to the target view's position, one from the nearest left and another from the nearest right. The two projected images are then blended to form a synthesised image. The pixel value in the synthesised image is created by blending the respective pixel values in the projected images with unequal weights, where the weights are inversely proportional to the distance from the target view. For fair comparison, this reference approach employs the same view discarding approach (i.e. priority method) as the proposed system and utilises MPEG VSRS to estimate discarding views.

In this experiment, only one intermediate view was discarded at a time. Average reconstruction quality, which was calculated in terms of PSNR, was estimated using only discarded views for each piece of MVV content.

Figure 8.9 shows average PSNR scores for the proposed view reconstruction and the MPEG VSRS algorithms for the *Book Arrival*, *Newspaper* and *Café* sequences. The transmission bitrate included the total transmitted data and SI. As seen in the RD curves, 2.8, 7.13 and 5.04 dB average PSNR gains (ΔP) are achieved over the entire range of bitrates with respect to the MPEG VSRS for *Book Arrival*, *News paper* and *Café* sequences, respectively.

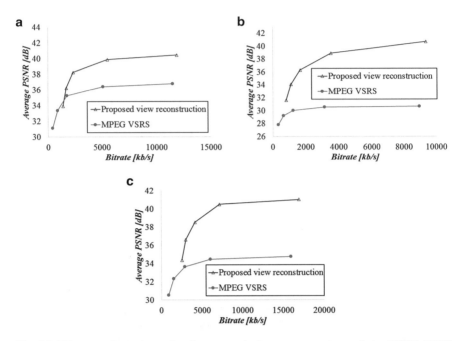

Fig. 8.9 RD comparison curves for the proposed view reconstruction and the MPEG VSRS algorithms. (**a**) *BookArrival*. (**b**) *Newspaper*. (**c**) *Café*

Experiment results demonstrate that the proposed view reconstruction algorithm presents significant coding efficiency compared to the MPEG VSRS method. The overall gains can be explained by two intrinsic novelties of the proposed reconstruction approach: (1) exploiting all available views (*see* Fig. 8.3) during the view reconstruction process. This feature provides high-quality occlusion filling, which enhances the overall view estimation quality at the decoder. High numbers of reference views minimise the occlusion problems on the estimated view and improve the view estimation performance. (2) optimisation process in the variable block-size selection. This process alternates between metadata overhead and view estimation quality, which allows the transmission of an optimum overhead of metadata without significantly decreasing the overall quality.

8.5.2.4   Effect of the Number of Discarded Views

Figure 8.10 depicts the average PSNR versus the total bitrate requirement for three different MVV contents. In this experiment, several views were discarded and then reconstructed using other available neighbouring views at the receiver side.

The reported results clearly demonstrate that a significant PSNR gain is achieved by the proposed approach against MPEG VSRS for all tested MVV content. This advantage is due to the high-quality view estimation and decreasing bitrate with SI.

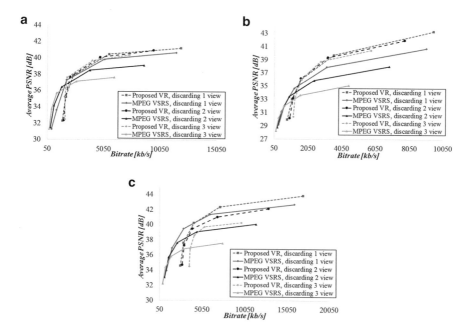

Fig. 8.10 Effect of the number of discarded views on the overall MVV streaming quality. (**a**) RD curves for *BookArrival*. (**b**) RD curves for *Newspaper*. (**c**) RD curves for *Café*

For the *Newspaper* sequence, the gain is more visible because of their high spatial complex scene (*see* SI value in Fig. 8.6). Moreover, the coding gain gap between the proposed approach and MPEG VSRS is smaller for the *Book Arrival* sequence relative to other sequences. The main reason is that both sequences contain complex object motion, which produces new occlusion areas on the projected views. This problem can be solved by transmitting a new codebook when the existing codebook is not sufficient for estimating new occlusion areas with optimum performance. Performance reduction also links to the accuracy of depth maps. The *Café* sequence, contains inaccurate depth maps in comparison to other sequences, demonstrates low performance below 5000 kbps. The reason is the proposed approach requires a greater number of weighting coefficients to reconstruct with erroneous depth maps with higher quality.

Furthermore, it is clearly observed that as the number of discarded views increases, PSNR values for each content severely decrease. The overall MVV streaming performance depends on the quality of the estimated views; hence, it is also linked to the accuracy of the depth maps and the efficiency of the view estimation algorithm. Obviously, it also depends on the complexity of the scene that needs to be estimated. This experiment clearly demonstrates that MPEG VSRS weaknesses affect adaptation performance. In addition, the results show that the proposed approach helps enhance adaptation performance by a notable margin consistently over a different number of discarded views, e.g. 1, 2 and 3.

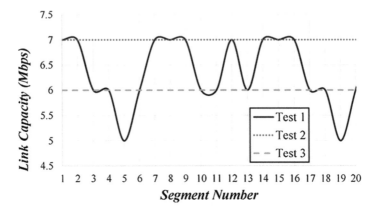

Fig. 8.11 Link capacity of the user in the network

8.5.2.5 Performance Analysis over the Dynamic Network Environment

In order to evaluate the impact of the adaptation pattern (i.e. view discarding order in the MPD), a test pattern with instantaneous network throughput changes was used as depicted in Fig. 8.11. `Test 1` corresponds to the varying link capacity case, whereas `Test 2` and `Test 3` correspond to the fixed link capacity case throughout the streaming. MVV temporal segments were selectively discarded based on the available bandwidth capacity.

In this experiment, the HEVC standard was used with the MPEG DASH. All streams were divided into segments, encoded with various QPs (*see* Sect. 8.5.1), and stored in the DASH server. Avoiding inter-view dependencies within the proposed framework will offer increased flexibility in view combinations that can be discarded independently and are replaced with the corresponding SI stream. Also, it prevents potential inter-view error propagation. In addition, the proposed adaptation system may reduce the quality of all the transmitted views based on the available bandwidth.

Furthermore, MPEG VSRS was used as an adaptation reference, in which MVV temporal segments were selectively discarded based on the available bandwidth. Also, this adaptation reference may also increases/decreases QP of the all transmitted views.

The view discarding order (i.e. which views were discarded) in `Test 1` for both adaptation methods is shown in Table 8.2.

In addition, Table 8.3 shows the comparison in terms of PSNR and subjective scores that are reported as an average of all views (delivered and discarded/estimated views). For subjective experiments, MOS values are converted to quality scales (bad, poor, fair, good and excellent) based on the ITU-R BT.500 recommendation.

The results depicted in the Table 8.3 show that the proposed adaptation method consistently outperforms the reference method, both objectively and subjectively, in all test conditions.

Table 8.2 View discarding pattern between first segment and seventh segment for varying link capacity test case

Segment range (0.5 s/segment)		1–2	2–3	3–4	4–5	5–6	6–7
Method	Content	Discarded viewpoint number(s)					
Proposed view recovery	Café	–	–	4	3	–	4
MPEG view synthesis		–	–	4	4	–	4
Proposed view recovery	Newspaper	5	4,3	3,5	3,4	4,5	5
MPEG view synthesis		3	5,3	5,3	5,3	5,3	5
Proposed view recovery	BookArrival	–	8	–	8	7	–
MPEG view synthesis		–	8	–	8	8	–

Table 8.3 Comparison of the adaptation methods

	Method	Quality	Test conditions		
			Test 1	Test 2	Test 3
Café	Proposed view recovery	PSNR (dB)	42.03	42.22	41.72
		Subjective	Excellent	Excellent	Excellent
	MPEG view synthesis	PSNR (dB)	41.59	42.22	41.28
		Subjective	Excellent	Excellent	Good
Newspaper	Proposed view recovery	PSNR (dB)	42.53	42.41	41.65
		Subjective	Excellent	Excellent	Excellent
	MPEG view synthesis	PSNR (dB)	40.52	41.17	39.18
		Subjective	Good	Good	Fair
BookArrival	Proposed view recovery	PSNR (dB)	39.37	39.74	39.68
		Subjective	Excellent	Excellent	Excellent
	MPEG view synthesis	PSNR (dB)	37.61	39.74	39.52
		Subjective	Good	Good	Good

8.5.2.6 Visual Quality Performance

The presented results demonstrate that the video coding standards with the proposed approach performs objectively (PSNR) better than the MPEG VSRS. These improvements are also visible in Fig. 8.12, which illustrates the visual quality of the *Book Arrival*, *News paper* and *Café* sequences.

In this analysis, thumbnails of the reconstructed discarded view images were captured and illustrated. Around the object edges, the occluded areas that do not exist with the proposed approach are clearly shown. It is clear that particular object boundaries that look distorted in the MPEG VSRS are better conserved with the proposed view reconstruction method. This is because boundary pixels is the estimated frames are not scattered, given that the local high frequency components are conserved successfully with an additional SI. Consequently, a sharper and more robust perception is achieved.

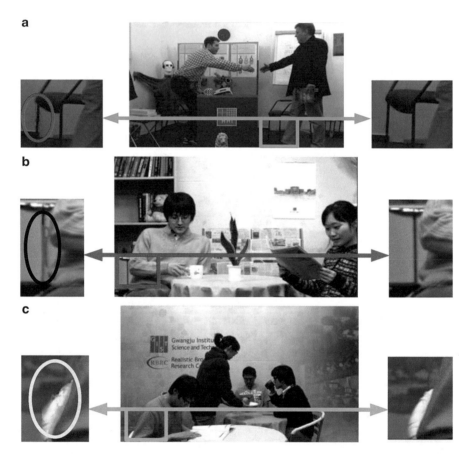

Fig. 8.12 Subjective reconstruction performance of the missing view between the MPEG VSRSR and the proposed approach. The resulting reconstruction views are shown for three different MVV contents: *Book Arrival*, *News paper* and *Café*. The most representative areas of distortion are marked. (**a**) Reconstruction view of *BookArrival*. (**b**) Reconstruction view of *Newspaper*. (**c**) Reconstruction view of *Café*

8.6 Conclusions

To maintain the perceived 3D MVV quality in congested networks, this chapter suggests a novel adaptive delivery scheme. The proposed method yields a superior performance over a wide range of channel conditions. In this streaming system, some views are discarded at times of network congestion in an intelligent way to maximise the resultant reconstruction performance on the end-user. The discarded views are reconstructed using only a small amount of additional metadata that is estimated in the server and sent to the receiver.

In the proposed method, the additional metadata is calculated using adjacent views in the server and delivered to the end-user at times of congestion in the network. Also, a novel view reconstruction method is designed to take into account the received side information (SI) for improved view recovery performance. In order to help facilitate a quality-aware bandwidth adaptation mechanism, the best sets of views to be discarded are calculated for various network throughput levels, such that the best overall MVV reconstruction quality is achieved on the end-user.

The proposed adaptive 3D MVV streaming method was evaluated using the HEVC and MPEG-DASH standards. The experiment results have shown that significant quality improvements are obtained under challenging network conditions. As a future study, perceptual 3D QoE metrics will be integrated in the adaptation strategy.

References

1. Adobe (2016) Adobe HTTP Dynamic Streaming (HDS). http://www.adobe.com/devnet/hds. html
2. Apple (2016) Apple HTTP live streaming. https://developer.apple.com/streaming/.
3. Benzie P, Watson J, Surman P, Rakkolainen I, Hopf K, Urey H, Sainov V, Kopylow CV (2007) A survey of 3DTV displays: techniques and technologies. IEEE Trans Circuits Syst Video Technol 17(11):1647–1658
4. Bjøtegaard G (2001) Calculation of average PSNR differences between RD-curves (vceg-m33). Technical Report M16090, VCEG Meeting (ITU-T SG16 Q.6), Austin
5. Dannewitz C (2009) Netinf: an information-centric design for the future Internet. In: Proceedings 3rd GI/ITG KuVS workshop on the future internet
6. De Simone F, Dufaux F (2013) Comparison of DASH adaptation strategies based on bitrate and quality signalling. In: 2013 IEEE 15th international workshop on multimedia signal processing (MMSP), pp 087–092
7. Feldmann I, Mueller M, Zilly F, Tanger R, Mueller K, Smolic A, Kauff P, Wiegand T (2008) HHI test material for 3D video. Technical Report MPEG2008/M15413, ISO/IEC JTC1/SC29/WG11, Archamps
8. Fielding R, Gettys J, Mogul J, Frystyk H, Masinter L, Leach P, Berners-Lee T (1999) Hypertext transfer protocol–http/1.1. Technical report
9. Giladi A, Stockhammer T (2015) Descriptions of Core Experiments on DASH Amendment. Technical Report MPEG2015/N15592, ISO/IEC JTC1/SC29/WG11/, Warsaw
10. Giladi A, Stockhammer T (2015) Technologies under consideration for dynamic adaptive streaming over http 23009, parts 1, 3 and 4. Technical Report MPEG2014/N15214, ISO/IEC JTC1/SC29/WG11, Geneva
11. Helle P, Oudin S, Bross B, Marpe D, Bici MO, Ugur K, Jung J, Clare G, Wiegand T (2012) Block merging for quadtree-based partitioning in HEVC. IEEE Trans Circuits Syst Video Technol 22(12):1720–1731
12. Ho Y-S, Lee E-K, Lee C (2008) Multiview video test sequence and camera parameters. Technical Report MPEG2008/M15419, ISO/IEC JTC1/SC29/WG11, Archamps
13. Information technology – Dynamic Adaptive Streaming over HTTP (DASH) – Part 1: Media presentation description and segment formats. Technical Report ISO/IEC 23009-1:2012, Geneva (2012)
14. ITU-R BT Recommendation (2002) 500-11, methodology for the subjective assessment of the quality of television pictures
15. Jacobson V (1988) Congestion avoidance and control. ACM SIGCOMM Comput Commun Rev 1:314–329

16. Jacobson V, Smetters DK, Thornton JD, Plass MF, Briggs NH, Braynard RL (2009) Networking named content. In: Proceedings of the 5th international conference on emerging networking experiments and technologies, CoNEXT '09. ACM, New York, pp 1–12
17. Kang Y-S, Lee E-K, Jung J-I, Lee J-H, Shin I-Y (2009) 3D video test sequence and camera parameters. Technical Report MPEG2009/M16949, ISO/IEC JTC1/SC29/WG11, Sian
18. Kanungo T, Mount DM, Netanyahu NS, Piatko CD, Silverman R, Wu AY (2002) An efficient k-means clustering algorithm: analysis and implementation. IEEE Trans Pattern Anal Mach Intell 24(7):881–892
19. Kondoz AM (2004) Front Matter, in Digital Speech: coding for low bitrate communication systems, Second Edition, John Wiley & Sons, Ltd, Chichester, UK.
20. Lederer S, Müller C, Timmerer C (2012) Dynamic adaptive streaming over HTTP dataset. In: Proceedings of the 3rd multimedia systems conference, MMSys '12. ACM, New York, pp 89–94
21. Lewandowski F, Paluszkiewicz M, Grajek T, Wegner K (2012) Subjective quality assessment methodology for 3D video compression technology. In: 2012 international conference on signals and electronic systems (ICSES), pp 1–5
22. Lightstone M, Mitra SK (1997) Quadtree optimization for image and video coding. J VLSI Signal Process Syst Signal Image Video Technol 17(2–3):215–224
23. Microsoft (2016) Microsoft Smooth-Streaming. http://www.iis.net/downloads/microsoft/smooth-streaming
24. Miller K, Quacchio E, Gennari G, Wolisz A (2012) Adaptation algorithm for adaptive streaming over HTTP. 2012 19th international packet video workshop (PV), pp 173–178
25. Moore DS (1976) Chi-square tests. Defense Technical Information Center
26. Müller C, Lederer S, Timmerer C (2012) An evaluation of dynamic adaptive streaming over HTTP in vehicular environments. In: Proceedings of the 4th workshop on mobile video, MoVid '12. ACM, New York, pp 37–42
27. Müller K, Merkle P, Wiegand T (2011) 3D video representation using depth maps. Proc IEEE 99(4):643–656
28. Müller K, Schwarz H, Marpe D, Bartnik C, Bosse S, Brust H, Hinz T, Lakshman H, Merkle P, Rhee FH, Tech G, Winken M, Wiegand T (2013) 3D high-efficiency video coding for multi-view video and depth data. IEEE Trans Image Process 22(9):3366–3378
29. Onural L (2007) Television in 3D: What are the prospects? Proc IEEE 95(6):1143–1145
30. Oyman O, Singh S (2012) Quality of experience for HTTP adaptive streaming services. IEEE Commun Mag 50(4):20–27
31. P.910 ITU-T Recommendation. Subjective video quality assessment methods for multimedia applications (1999)
32. Paul S, Yates R, Raychaudhuri D, Kurose J (2008) The cache-and-forward network architecture for efficient mobile content delivery services in the future Internet. In: Innovations in NGN: future network and services, 2008. K-INGN 2008. First ITU-T Kaleidoscope academic conference, pp 367–374
33. Rizzo L (1997) Dummynet: a simple approach to the evaluation of network protocols. ACM SIGCOMM Comput Commun Rev 27(1):31–41
34. Sanchez Y, Schierl T, Hellge C, Wiegand T, Hong D, De Vleeschauwer D, Van Leekwijck W, Le Louédec Y (2012) Efficient HTTP-based streaming using scalable video coding. Signal Process Image Commun 27(4):329–342
35. Seufert M, Egger S, Slanina M, Zinner T, Hobfeld T, Tran-Gia P (Firstquarter 2015) A survey on quality of experience of HTTP adaptive streaming. IEEE Communications Surveys Tutorials 17:469–492
36. Smolic A, Kauff P (2005) Interactive 3D video representation and coding technologies. Proc IEEE 93(1):98–110
37. Stockhammer T (2011) Dynamic adaptive streaming over HTTP: standards and design principles. In: Proceedings of the second annual ACM conference on multimedia systems, MMSys '11. ACM, New York, pp 133–144

38. Sugiyama Y (1986) An algorithm for solving discrete-time Wiener-Hopf equations based upon Euclid's algorithm. IEEE Trans Inf Theory 32(3):394–409
39. Sullivan GJ, Baker R (1994) Efficient quadtree coding of images and video. IEEE Trans Image Process 3(3):327–331
40. Sullivan GJ, Wiegand T (1998) Rate-distortion optimization for video compression. IEEE Signal Process Mag 15(6):74–90
41. Swaminathan V, Streeter K, Bouazizi I (2015) Working Draft for 23009-6: DASH over Full Duplex HTTP-based Protocols (FDH). Technical Report MPEG2015/N15532, ISO/IEC JTC1/SC29/WG11/, Warsaw
42. Tanimoto M, Fujii T, Suzuki K, Fukushima N, Mori Y (2008) Reference softwares for depth estimation and view synthesis. Technical Report MPEG2008/M15377, ISO/IEC JTC1/SC29/WG11, Archamps
43. Tanimoto M, Fujii M, Panahpour M, Wilderboer M (2009) Depth estimation reference software DERS 5.0. Technical Report MPEG2009/M16923, ISO/IEC JTC1/SC29/WG11, Xian
44. Tech G, Schwarz H, Müller K, Wiegand T (2012) 3D video coding using the synthesized view distortion change. In: Picture coding symposium (PCS), 2012, Krakow, pp 25–28
45. Thomas E (2015) Ce-sissi report (for consideration). Technical Report MPEG2015/M36598, ISO/IEC JTC1/SC29/WG11/, Warsaw
46. Thomas E, Brandenburg R (2015) Analysis of URI Signing extension for segmented content. Technical Report MPEG2015/m36435, ISO/IEC JTC1/SC29/WG11/, Warsaw
47. Thomas E, Champel ML, Begen AC (2015) Report on sand core experiment (for review). Technical Report MPEG2015/M36453, ISO/IEC JTC1/SC29/WG11/, Warsaw
48. Vetro A, Sodagar I (2011) Industry and standards the MPEG-DASH standard for multimedia streaming over the internet. IEEE Multimedia 18(4):62–67
49. Vetro A, Wiegand T, Sullivan GJ (2011) Overview of the stereo and multiview video coding extensions of the H.264/MPEG-4 AVC standard. Proc. IEEE 99(4):626–642
50. Wang X, Zhang S (2015) CAPCO CE: server-generated mosaic service use case and requirements. Technical Report MPEG2015/M36163, ISO/IEC JTC1/SC29/WG11/, Warsaw
51. Wang Z, Bovik AC, Sheikh HR, Simoncelli EP (2004) Image quality assessment: from error visibility to structural similarity. IEEE Trans Image Process 13(4):600–612
52. Wegner K, Stankiewicz O, Klimaszewski K, Domański M (2010) Comparison of multiview compression performance using MPEG-4 MVC and prospective HVC technology. Technical Report MPEG M17913, ISO/IEC JTC1/SC29/WG11, Geneve
53. Yaning L, Geurts J, Point J-C, Lederer S, Rainer B, Müller C, Timmerer C, Hellwagner H (2013) Dynamic adaptive streaming over CCN: a caching and overhead analysis. In: 2013 IEEE international conference on communications (ICC), pp 3629–3633
54. Zhao Y, Yu L (2010) A perceptual metric for evaluating quality of synthesized sequences in 3DV system. In: Proceedings of SPIE Vol, vol 7744, pp 77440X–1
55. Zhou C, Zhang X, Huo L, Guo Z (2012) A control-theoretic approach to rate adaptation for dynamic HTTP streaming. In: 2012 IEEE visual communications and image processing (VCIP), pp 1–6

Index

© Springer Science+Business Media New York 2017
A. Kondoz, T. Dagiuklas (eds.), *Connected Media in the Future Internet Era*,
DOI 10.1007/978-1-4939-4026-4

Printed in the United States
By Bookmasters